SpringerBriefs in Computer Science

SpringerBriefs present concise summaries of cutting-edge research and practical applications across a wide spectrum of fields. Featuring compact volumes of 50 to 125 pages, the series covers a range of content from professional to academic.

Typical topics might include:

- A timely report of state-of-the art analytical techniques
- A bridge between new research results, as published in journal articles, and a contextual literature review
- A snapshot of a hot or emerging topic
- An in-depth case study or clinical example
- A presentation of core concepts that students must understand in order to make independent contributions

Briefs allow authors to present their ideas and readers to absorb them with minimal time investment. Briefs will be published as part of Springer's eBook collection, with millions of users worldwide. In addition, Briefs will be available for individual print and electronic purchase. Briefs are characterized by fast, global electronic dissemination, standard publishing contracts, easy-to-use manuscript preparation and formatting guidelines, and expedited production schedules. We aim for publication 8–12 weeks after acceptance. Both solicited and unsolicited manuscripts are considered for publication in this series.

**Indexing: This series is indexed in Scopus, Ei-Compendex, and zbMATH **

Yu Zhou · Xiao Zhang · Sam Kwong

Computational Intelligence for High-Dimensional Machine Learning

A Feature Selection Perspective and Its Real-World Applications

 Springer

Yu Zhou
College of Computer Science and Software
Engineering
Shenzhen University
Shenzhen, Guangdong, China

Xiao Zhang
Department of Computer Science
South-Central Minzu University
Wuhan, Hubei, China

Sam Kwong
School of Data Science
Lingnan University
Tuen Mun, Hong Kong

ISSN 2191-5768   ISSN 2191-5776 (electronic)
SpringerBriefs in Computer Science
ISBN 978-981-96-2686-1   ISBN 978-981-96-2687-8 (eBook)
https://doi.org/10.1007/978-981-96-2687-8

This Springer imprint is published by the registered company Springer Nature Singapore Pte Ltd.
The registered company address is: 152 Beach Road, #21-01/04 Gateway East, Singapore 189721, Singapore

If disposing of this product, please recycle the paper.

Preface

Welcome to read *Computational Intelligence for High-Dimensional Machine Learning*! This book explores the intersection of computational intelligence techniques and advanced feature selection approaches for high dimensional data, presenting a comprehensive journey through the realms of evolutionary algorithms and deep neural networks based feature selection, global and local feature selection models, and their applications in the emerging real-world problem solving.

Organization of the Book:

- Chapter 1: Introduction of High Dimensional Machine Learning
  Sets the stage by introducing high dimensional machine learning and feature selection approaches. It provides an overview of the challenges and opportunities in the field.
- Chapter 2: Feature Selection and Computational Intelligence Methods
  Dives into the fundamental concepts of feature selection and computational intelligence techniques. It lays the groundwork for understanding how these methodologies contribute to advancements in feature selection.
- Chapter 3: Evolutionary Algorithm Based Global Feature Selection
  Explores the novel approach of treating global feature selection as a multiobjective optimization problem. The proposed problem-specific non-dominated sorting genetic algorithm is introduced, emphasizing its accuracy-preferred domination operator and quick bit mutation.
- Chapter 4: Evolutionary Algorithm Based Local Feature Selection
  Shifts the focus to local feature selection (LFS), proposing a region purity-based LFS that addresses the limitations of existing algorithms. The enhanced nondominated sorting genetic algorithm III, network-inspired crossover operator, and quick bit mutation contribute to the effectiveness of RP-LFS.
- Chapter 5: Deep Neural Network Based Hybrid Feature Selection
  Leverages the advancements in sparse group lasso mechanisms for feature selection. Introduces a hybrid attention-based deep neural network for embedded feature selection, which has been successfully applied in multi-sensor human activity recognition task for validation.

- Chapter 6: Real-World Case Study
  After discussing various advanced FS methods, we explore practical applications of computational intelligence based feature selection for high dimensional machine learning across diverse fields. The first subsection addresses fault diagnosis problems in industrial informatics, proposing an Improved Localized Feature Selection (LFS) method, based on Multi-Objective Binary Particle Swarm Optimization (LFS-MOBPSO). The second subsection tackles the classification of DNA microarray data in bioinformatics, which presents a Cooperative Coevolutionary Multiobjective Genetic Programming (CC-MOGP) approach to perform embedded FS and ensemble learning to avoid overfitting.
- Chapter 7: Conclusions
  Summarizes the key findings and contributions of the book. Explores potential future directions in the field of computational intelligence based feature selection for high dimensional machine learning.

This book is intended for researchers, practitioners, and students interested in the interdisciplinary field of machine learning and computational intelligence. It is designed to provide readers with a comprehensive understanding of high dimensional machine learning, optimization techniques, and the application of computational intelligence-based feature selection in real-world scenarios.

I extend my heartfelt gratitude to the contributors, researchers, and students whose efforts have been instrumental in bringing this book to fruition. Their commitment to advancing the fields of computational intelligence and high-dimensional machine learning has significantly enriched the insights and content presented in these pages.

A special note of thanks goes to my wife, Hannah, and my daughter, Amy. Their unwavering support and love have been my greatest source of strength and inspiration throughout this journey.

Shenzhen, China   Yu Zhou
November 2024

Contents

Chapter 1
Introduction of High Dimensional Machine Learning

Abstract The chapter introduces the growing challenges posed by high-dimensional data in machine learning applications and emphasizes the significance of feature selection (FS) to enhance classification performance. Key issues and challenges for high-dimensional machine learning are outlined, including the curse of dimensionality, overfitting, interpretability and scalability. The rationale of using computational intelligence based feature selection approaches to improve the scalability and interpretability of the learning model is concluded.

Keywords High dimensional data · Machine learning · Feature selection

1.1 Problem Definition

With the development of modern data sampling and acquisition techniques, machine-learning applications have witnessed a rise in the quantity of high-dimensional data. High-dimensional machine learning focuses on analyzing and modeling data where the number of features (dimensionality) is significantly larger than, or comparable to, the number of observations. This situation is frequently encountered in a variety of domains where complex and detailed datasets are common [1–3], such as genomics, where thousands of genes or genetic markers are analyzed simultaneously; human action recognition (HAR), which involves many informative features with different modalities; industrial informatics, where various industrial production systems rely on a lot parameters to be optimized and texts are represented using extensive vocabularies or embeddings; and financial modeling, which may include a vast array of financial indicators and metrics. While high-dimensional data presents opportunities for uncovering complex patterns, it also introduces unique challenges that require specialized techniques and methodologies.

In high-dimensional machine learning, the primary goal is to develop algorithms that can effectively learn from data with a high number of features (or predictors) relative to the number of samples. Formally, given a dataset $\{\mathbf{x}, \mathbf{y}\}$, where $\mathbf{x} \in R^{n \times p}$ denotes the matrix containing n samples and p features and $\mathbf{y} \in R^{n \times 1}$ denotes the vector consisting of the labels for n samples. For high-dimensional machine learning, with $p \gg n$, traditional statistical learning models suffer from overfitting and generalization problems and a number of issues and challenges are raised accordingly.

1.2 Key Issues and Challenges

There are majorly the following issues and challenges that need to be considered for learning high-dimensional data.

- **The Curse of Dimensionality**: It refers to the exponential increase in computational complexity and data sparsity as dimensionality grows. This results in poor generalization due to overfitting, increased storage and computational requirements and difficulty in estimating reliable statistical measures due to limited sample density.
- **Model Overfitting**: With $p \gg n$, traditional learning algorithms may fit the noise rather than the underlying data distribution, leading to poor out-of-sample performance. In addition, high-dimensional datasets often contain irrelevant or redundant features, which can obscure important patterns and increase the noise in models. Therefore, it is important to identify the most informative features to perform dimension reduction and achieve a more compact feature representation.
- **Interpretability**: Models trained on high-dimensional data can be complex, making it challenging to interpret the relationships between features and outcomes. On one hand, for the traditional approaches like Principal Component Analysis (PCA) [4] and t-SNE [5], aim to project data into lower-dimensional spaces; on the other, these transformations may lose critical domain-specific information. Especially for the deep learning based models, the relationship between data and labels is learned through a black-box manner via some embedded dimension reduction operations, such as variational autoencoder (VAE) [6], making it difficult to understand the importance of the features in the original space.
- **Scalability**: For real-world high-dimensional problems, algorithms must scale to millions of features, so, how to reduce the computational complexity and improve the efficiency of both training and inference stages is very critical. The combination with efficient parallelization and distributed computing can make the problem easier to solve. However, the relevance between the features may be ignored during this divide-and-conquer manner, still suffering from information loss or noisy effect.

1.3 Addressing High-Dimensional Challenges with Feature Selection

Feature selection (FS), a crucial preprocessing step that tries to improve classification performance by selecting the best subset of features from a large number of redundant features, is of great importance. Recent studies have shown that FS can reduce the dimensionality, avoid model overfitting and enhance the interpretability to some extent for traditional machine learning models [7–9]. When equipped with global search methods, FS can be be easily extended to a larger-scale problem, since identifying the feature subest can be essentially regarded as a search problem.

1.3.1 Categories of Feature Selection

In general, common FS approaches include: filter methods, which is based on statistical measures (e.g., correlation [10], mutual information [11]); Wrapper methods, which use predictive models to evaluate subsets of features; and Embedded methods, which incorporate feature selection as part of model training (e.g., LASSO [12] and some other regularization techniques, such as L2 regularization (Ridge regression) [13] and Elastic Net [14] (Combining L1 and L2 regularization for robustness).

1.3.2 Feature Selection as a Search Problem

FS involves two main processes: subset search and subset evaluation. Various methods, including sequential forward selection (SFS) [7], sequential backward selection (SBS) [15], and Plus-l take-away-r (PTA) [16], have been applied to address the challenge of searching for an optimal subset in high-dimensional datasets, which is inherently an NP-hard problem susceptible to local optima. Evolutionary algorithms (EAs), such as genetic algorithms (GA) [17], particle swarm optimization (PSO) [18], and differential evolution (DE) [19], are preferred due to their global search ability, offering better convergence and overcoming the limitations of traditional methods. Feature subset evaluation aims to establish criteria for assessing the chosen feature subset's goodness. Traditional criteria include training accuracy [20], often measured in classification error [21], and additional factors such as feature subset size [22], redundancy [23], relevance [24], or distance metric [25]. These criteria are combined with training error through weighted sums, introducing challenges in weight determination. Recent approaches focus on directly optimizing multiple objectives, ranging from two to four [26–28], yielding enhanced efficiency in subset selection and improved classification performance.

1.3.3 Advanced Feature Selection in Our Book

In this book, we mainly focus on the advanced FS methods that can improve the generalization ability, interpretability and scalability of the high dimensional learning model. Combining feature selection with a global search method can address the aforementioned scalability issue brought by high-dimensional data. In addition, incorporation of feature selection module into deep learning model can not only perform an efficient end-to-end learning for high-dimensional data, but also enhance the interpretability of the designed deep learning model.

Considering that computational intelligence (CI) methods such as evolutionary algorithms (EAs), have the global search ability, with proper algorithm and objective function design, more advanced CI algorithms, such as multiobjective optimization, multi-task optimization and multimodal optimization can be developed for FS. Therefore, EA-based FS methods are firstly discussed. Besides, artificial neural networks (ANN)- an important branch of CI, serve as the foundamentals for deep learning models. So, deep learning based FS, in the context of CI-based FS is also detailed in the following chapters.

References

1. Oguntala, G.A., et al.: SmartWall: novel RFID-enabled ambient human activity recognition using machine learning for unobtrusive health monitoring. IEEE Access **7**, 68022–68033 (2019)
2. Liang, J.-M., et al.: Smart interactive education system based on wearable devices. Sensors **19**(15), 3260 (2019)
3. Ordonez, F.J., et al.: In-home activity recognition: Bayesian inference for hidden Markov models. IEEE Pervasive Comput. **13**(3), 67–75 (2014)
4. Tipping, M.E., Bishop, C.M.: Probabilistic principal component analysis. J. R. Stat. Soc. Ser. B (Stat. Methodol.) **61**(3), 611–622 (1999). https://doi.org/10.1111/1467-9868.00194
5. van der Maaten, L., Hinton, G.: Visualizing data using t-SNE. J. Mach. Learn. Res. **9**(Nov), 2579–2605 (2008)
6. Kingma, D.P., Welling, M.: Auto-encoding variational Bayes (2013). arXiv:1312.6114
7. Whitney, A.W.: A direct method of nonparametric measurement selection. IEEE Trans. Comput. **100**(9), 1100–1103 (1971)
8. Kohavi, R., et al.: A study of cross-validation and bootstrap for accuracy estimation and model selection. In: Ijcai, vol. 14, no. 2, pp. 1137–1145. Montreal, Canada (1995)
9. Reunanen, J.: Overfitting in making comparisons between variable selection methods. J. Mach. Learn. Res. **3**, 1371–1382 (2003)
10. Yu, L., Liu, H.: Feature selection for high-dimensional data: a fast correlation-based filter solution. In: Proceedings of the 20th International Conference on Machine Learning (ICML-03) (2003)
11. Hancer, E., Xue, B., Zhang, M.: Differential evolution for filter feature selection based on information theory and feature ranking. Knowl. Based Syst. **140**, 103–119 (2018)
12. Simon, N., et al.: A sparse-group lasso. J. Comput. Graph. Stat. **22**(2), 231–245 (2013)
13. Corinna, C., Mohri, M., Rostamizadeh, A.: L2 regularization for learning kernels (2012). arxiv:1205.2653
14. Zou, H., Hastie, T.: Regularization and variable selection via the elastic net. J. R. Stat. Soc. Ser. B Stat Methodol. **67**(2), 301–320 (2005)
15. Marill, T., Green, D.: On the effectiveness of receptors in recognition systems. IEEE Trans. Inf. Theory **9**(1), 11–17 (1963)
16. Kudo, M., Sklansky, J.: Comparison of algorithms that select features for pattern classifiers. Pattern Recogn. **33**(1), 25–41 (2000)
17. De Stefano, C., Fontanella, F., Cristina Marrocco, A., Di Freca, S.: A GA-based feature selection approach with an application to handwritten character recognition. Pattern Recogn. Lett. **35**, 130–141 (2014)

18. Xue, B., Zhang, M., Browne, W.N.: Particle swarm optimization for feature selection in classification: a multi-objective approach. IEEE Trans. Cybern. **43**(6), 1656–1671 (2012)
19. Wang, P., Xue, B., Liang, J., Zhang, M.: Multiobjective differential evolution for feature selection in classification. IEEE Trans. Cybern. (2021)
20. John, G.H., Kohavi, R., Pfleger, K.: Irrelevant features and the subset selection problem. In: Machine Learning Proceedings 1994, pp. 121–129 (1994)
21. Jain, A., Zongker, D.: Feature selection: evaluation, application, and small sample performance. IEEE Trans. Pattern Anal. Mach. Intell. **19**(2), 153–158 (1997)
22. Chen, K., Xue, B., Zhang, M., Zhou, F.: An evolutionary multitasking-based feature selection method for high-dimensional classification. IEEE Trans. Cybern. (2020)
23. Wang, Z., Li, M., Li, J.: A multi-objective evolutionary algorithm for feature selection based on mutual information with a new redundancy measure. Inf. Sci. **307**, 73–88 (2015)
24. Tabakhi, S., Moradi, P.: Relevance-redundancy feature selection based on ant colony optimization. Pattern Recogn. **48**, 2798–2811 (2015)
25. Tran, B., Xue, B., Zhang, M.: Genetic programming for multiple-feature construction on high-dimensional classification. Pattern Recogn. **93**, 404–417 (2019)
26. Zaretalab, A., Hajipour, V., Sharifi, M., Shahriari, M.R.: A knowledge-based archive multi-objective simulated annealing algorithm to optimize series-parallel system with choice of redundancy strategies. Comput. Ind. Eng. (2015). https://doi.org/10.1016/j.cie.2014.11.008
27. Hancer, E., Xue, B., Zhang, M., Karaboga, D., Akay, B.: Pareto front feature selection based on artificial bee colony optimization. Inf. Sci. **422**, 462–479 (2018). https://doi.org/10.1016/j.ins.2017.09.028
28. Zhang, Y., Gong, D., Cheng, J.: Multi-objective particle swarm optimization approach for cost-based feature selection in classification. IEEE/ACM Trans. Comput. Biol. Bioinform. (TCBB), **14**(1), 64–75 (2017). (IEEE Computer Society Press)

Chapter 2
Feature Selection and Computational Intelligence Methods

Abstract This chapter discusses both global and local methods for feature selection, providing a brief overview of evolutionary algorithms and neural networks in computational intelligence. Global feature selection is currently the most widely applied method, leveraging its advantages such as global search capabilities, high adaptability, flexibility, and the ability to perform multi-objective optimization. Local feature selection introduces a novel approach, optimizing within local feature spaces, offering a convenient way to handle heterogeneous structured data and, to some extent, achieving better efficiency. Evolutionary algorithms, a category simulating biological evolution, find extensive applications in addressing complex optimization and search problems due to their global search capabilities and adaptability to intricate issues. Deep learning, characterized by constructing deep neural networks, facilitates automatic learning of abstract representations of complex data patterns. This enables the model to achieve significant performance improvements when dealing with large-scale datasets and complex tasks.

Keywords Computational intelligence · Evolutionary algorithms · Deep learning

2.1 Feature Selection Methods

2.1.1 Global Methods

Over the past few decades, the Feature Selection (FS) problem has been regarded as a multi-objective optimization problem (MOP), where the objective is to simultaneously minimize the number of features and maximize classification accuracy to obtain the corresponding feature subset. Multi-objective Genetic Algorithms (GAs) for FS generally aim to balance the relationships between different objectives [1]. In [2], NSGA-II [3] was employed to tackle the FS problem, resulting in higher accuracy and faster convergence. The researchers defined feature selection as a problem involving two competing objectives, attempting to find a set of optimal solutions, known as Pareto-optimal solutions, rather than a single optimal solution.

© The Author(s), under exclusive license to Springer Nature Singapore Pte Ltd. 2025
Y. Zhou et al., *Computational Intelligence for High-Dimensional Machine Learning*,
SpringerBriefs in Computer Science, https://doi.org/10.1007/978-981-96-2687-8_2

In [4], a new multi-objective genetic algorithm was developed to address filter-based FS, considering numerous novel objective functions for optimization. In [5], the consideration extended to the parameter settings of Support Vector Machines (SVM) and the selection of feature subsets. The proposed approach utilized NSGA-II and a multi-criteria selection strategy to obtain the optimal feature subset. SVM represents a powerful technique for pattern classification problems, yet its efficiency significantly depends on the choice of kernel function type and associated parameter values. This study treated FS as a multi-objective optimization problem, with classification accuracy, the number of support vectors, margin, and the selected features defining their objective functions. An optimization method based on the multi-objective genetic algorithm NSGA-II was proposed.

In [6], FS was conceptualized as a multi-objective optimization problem with three objectives to minimize. A non-dominated sorting genetic algorithm (PS-NSGA) tailored for specific problems was introduced. PS-NSGA incorporated a regional NSGA-III with a set of novel operators to obtain smaller feature subsets while maintaining competitive classification accuracy. This approach increased the likelihood of high-accuracy individuals surviving in the population, and PS-NSGA adopted fast bit mutation to overcome limitations of traditional bit-string mutation, enhancing efficiency. Furthermore, PS-NSGA designed mutation-retry and combination operators for faster and better convergence. Finally, a solution selection strategy was employed to determine the most suitable feature subset from the obtained Pareto solutions.

These algorithms, based on global feature space, showcase the advantages of using multi-objective genetic algorithms to address the FS problem. They demonstrate the ability to simultaneously optimize multiple objective functions, achieve more accurate classification, and exhibit strong adaptability and robustness. Through crossover and mutation, genetic algorithms can extensively explore the global feature space, identifying different feature combinations, facilitating the discovery of genuinely useful features in the solution space, and ultimately finding the global optimum.

2.1.2 Local Methods

In contrast to the global feature space, Local Feature Selection (LFS), as a pioneering approach, partitions the entire sample space into multiple local regions and selects a feature subset for each local region [7]. Each region in the LFS sample space is associated with its unique optimized feature set, which may differ in membership and size across the entire sample space. This allows the feature sets to adapt optimally to local variations in the sample space. Additionally, since LFS makes no assumptions about the underlying structure of the data, this method is insensitive to the distribution of data in the sample space and exhibits excellent robustness against overfitting. Studies indicate that LFS can achieve better classification performance compared to global feature selection.

In [8], researchers aimed to obtain better local feature subsets by setting a crossover reaction threshold, calculating the distances between the central sample and other samples, and minimizing the ratio of the sum of distances and the threshold. In [9], LFS was treated as a linear programming optimization problem, seeking to minimize the sum of distances between samples of the same class while maximizing the sum of distances between samples of different classes. However, the selected range of feature numbers still needs manual adjustment, which is generally challenging to determine directly.

In [10], an efficient filtering local feature selection algorithm based on artificial immune systems was proposed. The algorithm introduced a cloning selection algorithm to explore the search space of optimal feature subsets and used local clustering as an evaluation criterion. It merged the minimization of the sum of distances between samples with the same label and the maximization of distances between samples with different labels into a single objective. An artificial immune system approach [11] was employed for resolution. To mitigate the impact of remote antigens on affinity function during measuring sample distances, the algorithm introduced a hyperparameter σ as part of calculating distance weights, which needed adjustment through cross-validation for different datasets.

In summary, these LFS-based feature selection methods are more suitable for handling data with heterogeneous structures, adapting to optimization in different scenarios of sample space. As they only need to optimize local feature space, sometimes they can achieve faster computational efficiency than global methods. Moreover, research indicates that LFS methods are significantly more robust compared to their global counterparts, better handling noise and outliers in the data.

2.2 CI Methods

2.2.1 Evolutionary Algorithms

Evolutionary Algorithms (EAs) constitute a class of optimization algorithms based on biological evolutionary processes, employed to address complex optimization and search problems. These algorithms simulate mechanisms such as natural selection, genetic inheritance, crossover, and mutation observed in biological evolution. By evolving within a candidate solution set, they aim to find the optimal solution or an approximation thereof for the given problem. Evolutionary Algorithms possess strong global search capabilities, allowing them to explore potential optimal solutions across the entire solution space. This adaptability makes them highly effective for complex problems, showcasing their versatility.

Furthermore, Evolutionary Algorithms inherently exhibit parallelism and distribution, enabling parallel searches among multiple individuals, thus enhancing search efficiency. The stepwise optimization of solution sets through operations such as selection, crossover, and mutation demonstrates the flexibility and robustness of these algorithms, allowing them to adapt to diverse problems and environments. Evolutionary Algorithms find applications across various problem domains, including function optimization, combinatorial optimization, parameter tuning, and optimization of machine learning model parameters. Owing to their ability for global search and adaptability to complex problem landscapes, Evolutionary Algorithms have gained widespread utilization in addressing practical problems.

Over the past few decades, EAs have demonstrated potential in addressing Feature Selection (FS) problems. The pioneering work of Siedlecki and Sklansky highlighted that Genetic Algorithms (GAs) could achieve better performance than other traditional algorithms [12]. Subsequent research has confirmed the advantages of GAs in FS [13, 14]. Although GAs were originally designed to work with backpropagation networks as classifiers, studies have shown their effectiveness in the context of backpropagation networks when using nearest neighbor classifiers to evaluate feature sets. However, achieving good performance often requires careful handling of numerous parameters. A practical approach to overcoming this limitation is through the hybridization of GAs, incorporating domain-specific knowledge into the algorithm. In [15, 16], researchers proposed hybrid genetic algorithms for FS by embedding several local search operations into traditional GAs for improved performance. The tuning of these operations' parameters was parameterized, and their effectiveness and temporal requirements were analyzed and compared. The research indicated that hybrid GAs significantly enhanced final performance and achieved control over subset size. Compared to classical GAs, hybrid GAs exhibited better convergence.

In [17], a two-stage evolutionary algorithm for FS was introduced, applying a specially designed GA in the first stage to minimize prediction errors. This method combined advantages from various artificial intelligence technologies. Specifically, the Radial Basis Function (RBF) neural network architecture utilized its simple topology and the fast fuzzy mean training algorithm as a non-linear modeling tool. A multi-objective optimization method was employed in two stages to select suitable variables: the first stage used a specially designed genetic algorithm to minimize prediction errors for monitoring dataset, and the second stage used simulated annealing to reduce the number of explanatory variables.

For Particle Swarm Optimization (PSO), an improved binary PSO was proposed in [18], where if *gbest* (the best position in the global population) did not improve over several iterations, it was reset to zero, avoiding premature convergence and effectively reducing the number of features. In [19], an enhanced PSO-based FS algorithm was introduced, featuring an adaptive feature selection process that dynamically interpreted the correlation and dependency of features within the feature subset. In [20], this method adopted a new representation to narrow down the search space, encoding each feature based on obtained cut points for better evaluation of candidate solutions, significantly reducing the solution space and improving classification accuracy.

In [21], a variable-length encoded PSO was proposed, endowing particles with different and shorter lengths, defining a smaller search space to enhance the performance of the particle swarm. By arranging features in descending order of correlation, the algorithm allowed shorter-length particles to achieve better classification performance. Additionally, with the proposed length variation mechanism, the PSO algorithm could break out of local optima, further narrowing down the search space and focusing on searching smaller, more effective regions. These strategies enabled PSO to obtain better solutions in a shorter time. The majority of research and experiments are sufficient to demonstrate the wide-ranging potential application of EAs in the field of feature selection.

2.2.2 Deep Neural Networks

Deep learning is a machine learning approach that revolves around constructing deep neural networks to learn and represent complex patterns in data. A distinctive feature of deep learning is the incorporation of multiple hidden layers (depth) within the network, enabling the model to automatically learn abstract representations of data, resulting in significant performance improvements when dealing with large-scale, high-dimensional datasets. The key components of deep learning include neural networks, deep learning algorithms, and activation functions.

Due to the inclusion of multiple hidden layers, deep learning neural networks can progressively abstract and extract high-level features from data, making them well-suited for capturing the complex structures within the data. Additionally, the ability of deep learning to leverage large-scale training data contributes to its high generalization performance, allowing for flexible applications across various research domains. With multi-layered architecture, deep learning models can learn hierarchical representations, from low-level features to high-level semantics, greatly enhancing their expressiveness and performance. As computational power increases and big data support grows, deep learning continues to adapt quickly to various application scenarios, showing exceptional accuracy and robustness when working with large-scale datasets. Additionally, its scalability and flexibility make it a powerful tool for solving real-world problems, continually pushing the boundaries of artificial intelligence.

In image classification tasks, scene images are typically first represented as feature vectors, which are then input into learning algorithms for training and testing. In feature-based image representation methods, feature extraction and feature encoding are fundamental steps, while feature selection plays another critical role. The goal of feature selection is to identify the most discriminative features from a large set of raw features, eliminating redundant or irrelevant ones to improve the efficiency and accuracy of the classification model. As a result, an increasing number of studies have applied deep neural networks to the feature selection problem [22–24]. With their powerful learning capabilities and automatic feature optimization mechanisms, deep neural networks can effectively extract representative and discriminative features from data. This approach not only addresses the limitations of traditional feature selection methods but also provides a more precise solution for tasks such as image classification.

Ribeiro and Lundberg et al. introduced LIME [25] and SHAP [26], which are inspired by feature weighting methods in interpretable machine learning (IML). These methods use feature weighting networks to preserve the physical meaning of hand-crafted features through feature selection (FS), thereby enhancing the model's interpretability at the feature level. By combining feature selection with interpretability, these approaches not only optimize model performance but also improve transparency and explainability. This is particularly valuable in applications where understanding and validating the model's decision-making process is crucial. Zhou et al. proposed the dfLasso-Net neural network [27], which incorporates a two-level weight computation module (TLWCM) consisting of sensor weight networks and feature weight networks. This module evaluates the importance of both sensors and features. Experimental results demonstrate that dfLasso-Net, as an effective filter-based feature selection method, offers greater flexibility.

Additionally, Zhou et al. [28] extended the sparse group Lasso mechanism to human activity recognition tasks, proposing a hybrid attention-based multi-sensor pruning and feature selection deep neural network—HAP-DNN. It achieves human activity recognition through feature selection via multi-sensor channel pruning. Studies indicate that deep learning is often a highly effective method, particularly when dealing with large-scale data and complex tasks.

References

1. Lac, H.C., Stacey, D.A.: Feature subset selection via multi-objective genetic algorithm. In: Proceedings. 2005 IEEE International Joint Conference on Neural Networks, 2005, vol. 3. IEEE (2005)
2. Hamdani, T.M., et al.: Multi-objective feature selection with NSGA II. In: Adaptive and Natural Computing Algorithms: 8th International Conference, ICANNGA 2007, Warsaw, Poland, 11–14 Apr 2007, Proceedings, Part I 8. Springer Berlin Heidelberg (2007)
3. Deb, K., et al.: A fast and elitist multiobjective genetic algorithm: NSGA-II. IEEE Trans. Evolut. Comput. **6**(2), 182–197 (2002)
4. Spolaôr, N., Lorena, A.C., Lee, H.D.: Multi-objective genetic algorithm evaluation in feature selection. In: Evolutionary Multi-Criterion Optimization: 6th International Conference, EMO 2011, Ouro Preto, Brazil, 5–8 Apr: Proceedings, vol. 6, p. 2011. Springer, Berlin Heidelberg (2011)
5. Bouraoui, A., Jamoussi, S., BenAyed, Y.: A multi-objective genetic algorithm for simultaneous model and feature selection for support vector machines. Artif. Intell. Rev. **50**, 261–281 (2018)
6. Zhou, Y., Zhang, W., Kang, J., Zhang, X., Wang, X.: A problem-specific non-dominated sorting genetic algorithm for supervised feature selection. Inf. Sci. **547**, 841–859 (2021)
7. Armanfard, N., Reilly, J.P., Komeili, M.: Local feature selection for data classification. IEEE Trans. Pattern Anal. Mach. Intell. **38**, 1217–1227 (2015)
8. Dudek, G.: An artificial immune system for classification with local feature selection. IEEE Trans. Evol. Comput. **16**, 847–860 (2012)
9. Armanfard, N., Reilly, J.P., Komeili, M.: Logistic localized modeling of the sample space for feature selection and classification. IEEE Trans. Neural Netw. Learn. Syst. **29**, 1396–1413 (2017)
10. Wang, Y., Li, T.: Local feature selection based on artificial immune system for classification. Appl. Soft Comput. **87**, 105989 (2020)
11. Farmer, J.D., Packard, N.H., Perelson, A.S.: The immune system, adaptation, and machine learning. Phys. D Nonlinear Phenomena **22**(1–3), 187–204 (1986)
12. Siedlecki, W., Sklansky, J.: A note on genetic algorithms for large-scale feature selection. In: Handbook of Pattern Recognition and Computer Vision, pp. 88–107, World Scientific (1993)
13. Brill, F.Z., Brown, D.E., Martin, W.N.: Fast generic selection of features for neural network classifiers. IEEE Trans. Neural Netw. **3**(2), 324–328 (1992)
14. Yang, J., Honavar, V.: Feature subset selection using a genetic algorithm. In: Feature Extraction, Construction and Selection, pp. 117–136. Springer (1998)

15. Oh, I.-S., Lee, J.-S., Moon, B.-R.: Local search-embedded genetic algorithms for feature selection. In: Object Recognition Supported by User Interaction for Service Robots, vol. 2, pp. 148–151 (2002)

16. Il-Seok, O., Lee, J.-S., Moon, B.-R.: Hybrid genetic algorithms for feature selection. IEEE Trans. Pattern Anal. Mach. Intell. **26**(11), 1424–1437 (2004)

17. Alexandridis, A., Patrinos, P., Sarimveis, H., Tsekouras, G.: A two-stage evolutionary algorithm for variable selection in the development of RBF neural network models. Chemom. Intell. Lab. Syst. **75**(2), 149–162 (2005)

18. Chuang, L.-Y., Chang, H.-W., Tu, C.-J., Yang, C.-H.: Improved binary PSO for feature selection using gene expression data. Comput. Biol. Chem. **32**(1), 29–38 (2008)

19. Unler, A., Murat, A.: A discrete particle swarm optimization method for feature selection in binary classification problems. Eur. J. Oper. Res. **206**(3), 528–539 (2010)

20. Tran, B., Xue, B., Zhang, M.: A new representation in PSO for discretization-based feature selection. IEEE Trans. Cybern. **48**(6), 1733–1746 (2017). (IEEE)

21. Tran, B., Xue, B., Zhang, M.: Variable-length particle swarm optimization for feature selection on high-dimensional classification. IEEE Trans. Evol. Comput. **23**(3), 473–487 (2019)

22. Camps-Valls, G., Mooij, J., Scholkopf, B.: Remote sensing feature selection by kernel dependence measures. IEEE Geosci. Remote Sens. Lett. **7**(3), 587–591 (2010)

23. Pal, M., Foody, G.M.: Feature selection for classification of hyperspectral data by SVM. IEEE Trans. Geosci. Remote Sens. **48**(5), 2297–2307 (2010)

24. Zou, Q., et al.: Deep learning based feature selection for remote sensing scene classification. IEEE Geosci. Remote Sens. Lett. **12**(11), 2321–2325 (2015)

25. Ribeiro, M.T., Sameer S., Guestrin, C.: "Why should i trust you?" Explaining the predictions of any classifier. In: Proceedings of the 22nd ACM SIGKDD International Conference on Knowledge Discovery and Data Mining (2016)

26. Lundberg, S.: A unified approach to interpreting model predictions (2017). arxiv:1705.07874

27. Zhou, Y., et al.: Energy-efficient and interpretable multisensor human activity recognition via deep fused lasso net. IEEE Trans. Emerg. Top. Comput. Intell. (2024)

28. Zhou, Y., et al.: A hybrid attention-based deep neural network for simultaneous multi-sensor pruning and human activity recognition. IEEE Internet of Things J. **9**(24), 25363–25372 (2022)

Chapter 3
Evolutionary Algorithm Based Global Feature Selection

Abstract Feature selection (FS) plays a crucial role in classification tasks. In this chapter, we consider feature selection as a multi-objective optimization problem and propose a Problem-Specific Non-dominated Sorting Genetic Algorithm (PS-NSGA) (Zhou et al. in Inf Sci (2021), [1]). In PS-NSGA, we have designed an accuracy-prioritized dominance operator, which increases the survival chances of individuals with higher classification accuracy in the population. Concurrently, a mutation retry operator and a combination operator were also designed to enable our algorithm to converge faster and more effectively. Finally, we developed a solution selection strategy to identify the most suitable feature subset. We conducted a series of experiments on 10 real-world high-dimensional datasets. The results demonstrate that PS-NSGA demonstrates outstanding performance in reducing the number of features while maintaining competitive classification accuracy, compared to some of the state-of-the-art evolutionary and traditional FS algorithms.

Keywords Evolutionary algorithm · Global feature selection · Multi-objective optimization

3.1 Objectives

In big data era that data dimensions are rapidly increasing, datasets often accumulate many redundant and irrelevant features. This can lead to a sharp increase in computational costs and a decrease in the classification accuracy of models [2]. The goal of Feature Selection is to address these issues. By choosing a subset of relevant features, feature selection can help mitigate overfitting and enhance classification performance.

Feature selection typically involves two main processes: subset search and subset evaluation [3]. Up to now, various subset search strategies have been applied to feature selection, such as Sequential Forward Selection (SFS) [4], sequential backward selection (SBS) [5], and Plus-l take-away-r (PTA) [6]. However, searching for the optimal feature subset from a feature pool is inherently a challenging NP-hard problem, Traditional methods are prone to getting stuck in local optima, especially

© The Author(s), under exclusive license to Springer Nature Singapore Pte Ltd. 2025   15
Y. Zhou et al., *Computational Intelligence for High-Dimensional Machine Learning*,
SpringerBriefs in Computer Science, https://doi.org/10.1007/978-981-96-2687-8_3

for high-dimensional data set. As evolutionary algorithms (EAs) have advantages in global search ability, many researchers have successfully attempted to apply Evolutionary Algorithms (EA) to feature selection in the past decade, examples include genetic algorithm (GA) [7], particle swarm optimization (PSO) [8, 9], differential evolution (DE) [10] and other bio-inspired search methods [11, 12].

In Feature Selection (FS), besides subset search, the evaluation indicator is also crucial for assessing the effectiveness of candidate solutions. Traditional methods often transform the problem into a single-objective optimization by linearly combining two metrics. However, determining the correct weight parameters, especially when dealing with multiple metrics, remains a challenge. For example, in [9], the training accuracy and the distance metric proposed in [13] are combined into one single objective. In recent years,researchers have shifted towards treating feature selection as a multi-objective optimization problem, aiming to optimize two or more objectives simultaneously. Various Multi-Objective Evolutionary Algorithms (MOEAs), such as Multi-Objective Simulated Annealing (SA) [14], multi-objective PSO [8], multi-objective GA [15] and multi-objective artificial bee colony [16], have been proposed to tackle this problem. These methods not only yield multiple Pareto-optimal solutions but also enhance decision flexibility, making feature selection more adaptable to complex environments.

Recently, the non-dominated sorting genetic algorithm, NSGA-III [17], has gained widespread attention for its excellent convergence and diversity in tackling multi-objective optimization problems, especially those with two or more objectives. This algorithm combines the advantages of non-dominated sorting schemes and decomposition-based methods. In addressing the issue where some regions near reference points cannot be associated with any individuals, an improved NSGA-III is proposed in [18], introducing a unified pool reservation strategy based on reference points. Additionally, in [19], an adaptive mutation operator was introduced to replace the fixed mutation rate in the original NSGA-III to optimize performance. In [20], an adaptive mechanism is proposed for adjusting ineffective reference points based on the distribution of individuals. Moreover, to cater to decision-makers' preferences for specific areas of the Pareto frontier, [21]introduced a method based on generating preferred reference points from user-provided aspiration points, integrating preference information into NSGA-III. These advancements inspire us to explore the application of NSGA-III in the field of feature selection, where making the design a problem-specific NSGA-III for FS is desirable.

Based on the earlier discussions, we propose a multi-objective optimization (MOP) problem for feature selection, aiming to minimize feature subset size, classification error rate, and a distance metric. To effectively address this MOP problem, we developed a feature-selection-specific version based on the NSGA-III framework. To better maintain population diversity in this specialized version of NSGA-III, we implemented a "precision-first" dominance strategy aimed at eliminating individuals with lower training accuracy. Additionally, we designed a rapid bit mutation strategy to overcome limitations associated with relying solely on random probability. Furthermore, to further explore the solution space, we introduced a dedicated mutation retry operator and a combination operator. In the final stage, when selecting suitable solutions, we considered both training accuracy and distance metrics. This approach aimed to choose solutions with higher accuracy.

3.2 Algorithm Design

3.2.1 Chromosome Encoding

In Genetic Algorithms (GA), chromosome encoding is a crucial concept, each chromosome represents a candidate solution to the problem and is generally represented as a vector. In our approach, we will use binary numbers to represent each gene on the chromosome, where "1' indicates the selection of a feature, and "0' denotes the non-selection of a feature. Taking a dataset with D dimensions as an example, the ith chromosome in the population is encoded by a D-bit string as follows:

$$X_i(t) = (x_{i,1}, x_{i,2}, \ldots, x_{i,j})$$
$$x_{i,j} \in \{0, 1\}, j = 1, 2, \ldots, D, i = 1, 2, \ldots, N \tag{3.1}$$

where $X_i(t)$ is the ith chromosome in the tth generation, N represets the population size, and $x(i, j) \in \{0, 1\}$ denotes the encoding of the jth gene on the ith chromosome.

3.2.2 Fitness Function

Fitness is a crucial metric used to evaluate the goodness of individuals. In our method, the fitness function comprehensively considers both classification error rate and distance metrics, thereby integrating the advantages of wrapper and filter methods. Additionally, the proportion of selected features is introduced as a third optimization objective, aiming to achieve more efficient dimensionality reduction.

3.2.2.1 Classification Error Rate

The classification error rate is calculated based on the performance of a specific classifier on a dataset. To address the issue of data imbalance in high-dimensional datasets, we employ the balanced error rate [22] for evaluation. The calculation is as follows:

$$balanced\_acc = \frac{1}{c} \sum_{i=1}^{c} TPR_i$$
$$balanced\_err = 1 - balanced\_acc \tag{3.2}$$

where c is the number of classes in dataset, TPR_i denotes the true positive ratio or the proportion of correctly predicted samples in class i. In our experiment, the weight for each class is set to $1/c$ and we used K-nearest neighbor (KNN) as the classifier to calculate true positive ratio for each class.

3.2.2.2 Distance Metric

Since the classification error rate is calculated by a specific classifier, it means that while we can obtain high-accuracy solutions on this classifier, there is no guarantee that the same effectiveness will be achieved on other classifiers with different decision boundaries. Additionally, a potential issue is the risk of overfitting in the classifier, leading to less reliable accuracy. To overcome these challenges,some evaluation metrics used in the filtering method are required, such as information gain, distance metric, Pearson correlation, and so on.

In this chapter, we have employed distance metrics [13] as the second objective in multi-objective optimization, aiming to reflect similarity for enhanced generalization when applied to various classifiers. Given that smaller distances indicate higher similarity between sample points, we achieve maximization of the distance between samples from different classes(Db) and minimization of the distance between samples within the same class (Dw), by minimizing the objective value 'distance'. The specific calculation is as follows:

$$D_b = \frac{1}{M} \sum_{i=1}^{M} \min_{\{j|j\neq i, class(S_i)\neq class(S_j)\}} Dis_{(S_i,S_j)} \tag{3.3}$$

$$D_w = \frac{1}{M} \sum_{i=1}^{M} \max_{\{j|j\neq i, class(S_i)=class(S_j)\}} Dis_{(S_i,S_j)} \tag{3.4}$$

$$distance = \frac{1}{1 + exp^{-5(D_w - D_b)}} \tag{3.5}$$

where M represents the number of samples in the training dataset, and $Dis(S_i, S_j)$ denotes the distance between two samples, S_i and S_j. This distance is measured using common metrics like Euclidean or Manhattan distance. In our PS-NSGA framework, we opt for the Manhattan distance, which is more appropriate for high-dimensional data scenarios. The inclusion of –5 in Eq. (3.5) aims to adjust the effective range of the sigmoid function. For additional insights into the exponent's power component, readers are referred to [13].By modifying the sigmoid function, the training data is scaled to the range of [0, 1].

3.2.2.3 Proportion of the Selected Features

To achieve better dimensionality reduction, we utilize the selected feature proportion as the third metric, described specifically as follows:

$$feature\_proportion = \frac{number\ of\ the\ selected\ features}{D} \tag{3.6}$$

where D denotes the number of features in the original dataset.

3.2.3 Accuracy-Preferred Domination

During the feature selection process, as we consider the proportion of selected features as an optimization objective, with the feature proportion decreasing, more individuals with a lower proportion of features but higher classification error rates are retained in the later stages of evolution, leading to a decline in average accuracy. In the later stages of training, a significant number of such individuals exist in the population, severely affecting the evolution of the population. Given this challenge, we have adjusted the traditional dominance principle to better suit the requirements of multi-objective feature selection algorithms. In our framework, we particularly emphasize the objective of classification accuracy. Therefore, we introduce the concept of "accuracy-preferred domination", defined as follows:

For two individuals x and y, let their objective values be represented as Obj_x, Obj_y, and their error rates as err_x and err_y, respectively. For a given tolerance of accuracy $\delta \in [0, 1]$, if any of the following conditions is satisfied, we say x dominates y in an accuracy-preferred manner:

$$\begin{aligned} err_y - err_x &\geq \delta \\ Obj_x &\prec Obj_y \end{aligned} \tag{3.7}$$

where $\prec$ denotes the usual domination. With accuracy-preferred domination, individuals exhibiting higher error rates are more likely to be eliminated, allowing individuals with superior accuracy to be retained in the population. This helps reduce the risk of getting stuck in local optima.

3.2.4 Genetic Operators

3.2.4.1 Quick Bit Mutation

In bit string mutation, certain genes are selected with a specific probability μ and their states are altered. For FS, applying this basic bit string mutation strategy alone can result in a slow reduction of the feature set, maintaining its size at a relatively balanced level. Specifically, when the selected feature ratio is low, genes encoded as 0 dominate the chromosome. Consequently, the majority of randomly selected genes for mutation are 0s, leading to an increase rather than a decrease in the size of the feature set. This slows down the rate of feature reduction, thereby decelerating the overall pace of the feature selection or evolutionary process.

To overcome this challenge, we have proposed a quick bit mutation, of which the process can be described as follows:

(1) Given a mutation rate μ, use (3.8) to calculate the number of bits (genes) to flip.

$$G = \mu \times min(count(0), count(1)) \tag{3.8}$$

(2) Randomly choose a set of G genes, which are either encoded as 1 or as 0.
(3) Alter the values of the selected genes.

The specific example is graphically represented in Fig. 3.1. Assume a mutation rate of 0.5 and a chromosome with four genes encoded as 1 and six genes encoded as 0. the number of genes G to be altered can be calculated as $0.5 \times min(4, 6) = 2$. Then, we select two genes randomly encoded as 1 (or 0) in the chromosome to be flipped.

The quick bit mutation strategy intensifies the selection pressure on genes within the chromosome. This is because genes encoded as 1 are highly likely to be selected for flipping, even if the majority are encoded as 0. Consequently, only those genes encoded as 1, representing the most critical features, stand a greater chance of persisting through the mutation process.

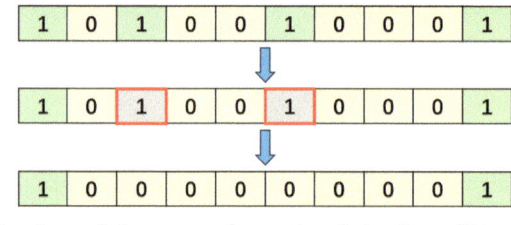

if two random 1-encoded genes are chosen, then their values will be changed to 0.

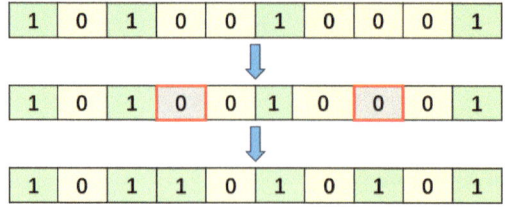

if two random 0-encoded genes are chosen, the values will be flipped to 1.

Fig. 3.1 quick bit mutation. The number of genes to flip bit G is $0.5 \times min(4, 6) = 2$

Choosing an appropriate mutation rate μ during the mutation process is crucial, as both too large and too small mutation rates μ can adversely affect the algorithm's performance. If the mutation rate μ is too small, the mutation speed will be too slow, significantly impacting the efficiency and exploration ability of the algorithm. Conversely, if the mutation rate μ is too large, it will easily generate an individual with fitness much better than the others, leading the algorithm into a local optimum without exploring more potential solutions.

3.2.4.2 Mutation-Retry Operator

If the mutation rate is too large, the differences in fitness among individuals in the population will be also large. Prominent individuals and their offspring may swiftly eliminate other individuals, occupy the population and control the direction of evolution. This is detrimental to maintaining population diversity, and it is prone to premature convergence and falling into local optima.

We introduce a mutation-retry mechanism to mitigate the negative effects mentioned above. During non-dominated sorting, different levels of non-dominated solution (denoted as $F_1, F_2, \ldots, F_l$) are generated. If the newly mutated individual is dominated by any individual in the last layer, it will inevitably be eliminated in the subsequent elitist selection, rendering all subsequent operations on this individual futile. Therefore, during the mutation process, we allow the new individual to be compared with individuals from the last layer in the previous non-dominated sorting. If the individual is dominated, there are k chances for mutation retry. The mutation-retry operation aims to randomly increase the number of mutations, narrowing the fitness disparity among individuals without slowing down the process of feature reduction.

3.2.4.3 Crossover Operator

In PS-NSGA, we have employed a uniform crossover operator. In uniform crossover, each gene in the offspring chromosome is independently selected from the corresponding positions in the two parental chromosomes. This selection mechanism does not favor either parent, resulting in the offspring inheriting approximately half of its genes from each parent. Unlike conventional crossover algorithms that generate two complementary offspring, our strategy selects only one offspring for further evolution.

Considering two parent chromosomes, $x1$ and $x2$, each with a length of 10, for each gene, we randomly choose the gene value in $x1$ or $x2$ for the offspring chromosome. This method is depicted in Fig. 3.2.

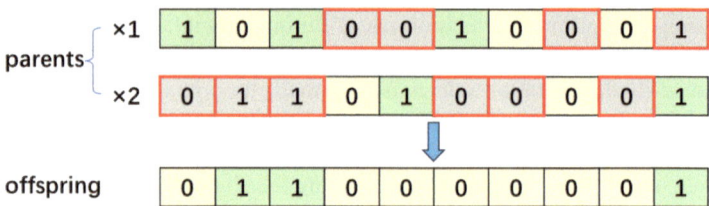

Fig. 3.2 The illustration of crossover operator. In ordinary crossover operator, there are two complementary offspring generated, but only one of which is chosen in our method

3.2.4.4 Combination Operator

The crossover operator's goal is to produce potential better individuals from the existing population. In feature selection (FS), employing uniform crossover can lead to the loss of certain parental features, as visually depicted in Fig. 3.2. However, during the advanced stages of evolution, it's likely that most or even all of the features in the parent chromosomes are crucial and should be retained. Therefore, we introduce a combination operator. This operator incorporates all features from both parents, enabling a more complete exploration in the later evolution stages. It's important to note that this combination operator is just a complement to the crossover operator. In our approach, only a few pairs of parents will perform it (10% of the total crossovers) while the remainder continues with uniform crossover.

3.2.5 Algorithm Framework

The multi-objective optimization algorithm PS-NSGA introduced in this chapter is based on NSGA-III, NSGA-III is distinguished by its ability to maintain population diversity through the use of a set of reference points. Particularly, NSGA-III has shown to be more effective than NSGA-II, in handling problems with more than two objectives, as evidenced by findings in [17].

3.2.5.1 Reference Point

In NSGA-III, diversity of obtained solutions is maintained by using a set of provided or predefined reference points. These reference points are placed on a normalized hyperplane situated in an M-dimensional objective space, where M represents the number of objectives. If each objective axis is partitioned into p divisions, then the total number of reference points (H) is determined by the following equation.

$$H = \binom{M+p-1}{p} \tag{3.9}$$

For a three-objective optimization problem with three divisions along each objective axis, the total number of reference points is $\binom{3+3-1}{3} = 10$. These reference points are uniformly distributed on a triangle defined by the vertices at $(1, 0, 0)$, $(0, 1, 0)$, and $(0, 0, 1)$, as illustrated in Fig. 3.3. The algorithm aims to approximate the Pareto front. Since the reference points are positioned on the normalized hyperplane, the resulting solutions can also achieve a uniform distribution on another hyperplane using the approaches based on reference points [17].

3.2.5.2 Algorithm Procedures

The procedures of the PS-NSGA are illustrated in Fig. 3.4, where N is the population size and T_{max} is the maximum number of iterations. The first stage is initialization, which randomly generates N chromosomes and creates H reference points. Then, we iterate through the following stages until the maximum number of iterations is reached:

(1) Genetic operators. Consider the population at generation t, denoted as P_t. Initially, mutation and crossover operations are applied to P_t, resulting in the creation of the offspring Q_t. Then, Q_t and P_t are combined to form a new population R_t.

(2) Non-dominated Sorting. The aim of non-dominated sorting is to sort the new population R_t into various non-dominated level$(F_1, F_2, \ldots)$, where individuals in a previous level dominate all individuals from the following levels, and no individual within the same level dominates another. Therefore, lower level represent better solutions, and the first level being the approximate Pareto Front (PF).

(3) Normalization of the population. The objective of normalization is to adjust the objective values of all individuals in the population to the proximity of the hyperplane where the reference points are distributed. Given that the range of each objective may differ, conventional L2 normalization is not suitable. For

Fig. 3.3 Ten evenly-distributed reference points for a three-objective problem with divisions=3

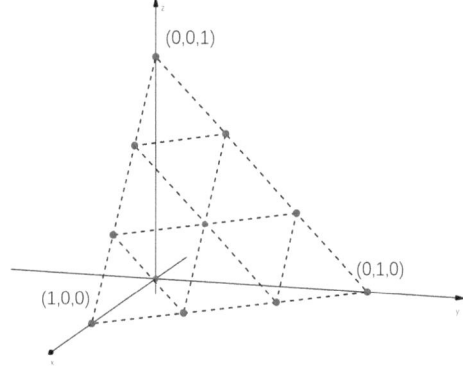

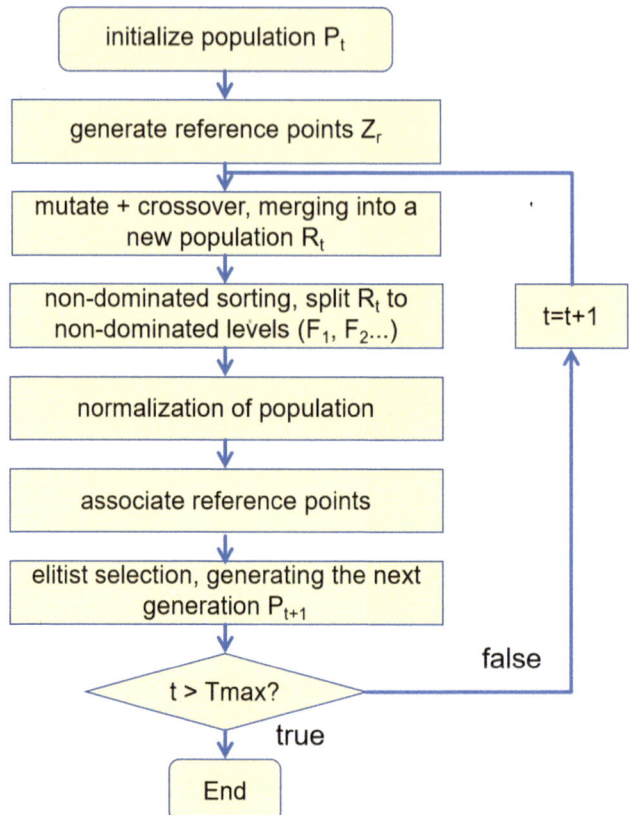

Fig. 3.4 algorithm procedures

normalization, an ideal point, an ideal point with the minimum value of each objective in the population is employed to translate all individuals. Subsequently, the extreme points are utilized to normalize the translated objectives. The detailed procedures are outlined in [17].

(4) Association with reference points. This process involves abstracting a ray from the origin to each reference point, as illustrated in Fig. 3.5. Each individual in the population is then associated with the reference point whose corresponding ray has the closest distance to the objective value of the individual. It is important to note that while each individual is associated with only one reference point, a single reference point may be associated with none or several individuals.

(5) Elitist Selection. After non-dominated sorting, the individuals are included in P_{t+1}, level by level, until adding a non-dominated level F_l causes $P_{t+1} > N$. If $|P_{t+1}| = N$, then P_{t+1} is established as the next generation, and the process moves to the next cycle; otherwise, additional K individuals are selected from F_l, where $K = N - |P_{t+1}|$. The methodology for selecting these K individuals

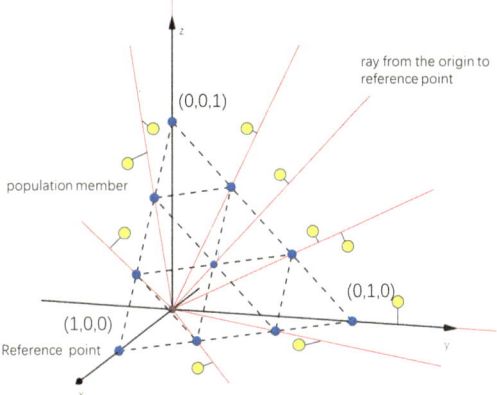

Fig. 3.5 associating population members with corresponding rays of reference points

from F_l includes several steps: initially, each reference point's count of associated individuals in P_{t+1} is computed. Subsequently, the reference point with
the minimum number of associations and with at least one individual associated
from F_l, is identified, and its nearest associated individual is incorporated into
P_{t+1}. This process is repeated K times until the total count of individuals P_{t+1}
satisfy $P_{t+1} = N$.

(6) If the maximum number of iterations T_{max} is reached, the genetic algorithm
(GA) is terminated, and the process proceeds to solution selection. Otherwise,
the iteration count t is incremented by 1, and the next iteration is executed.

3.2.5.3 Solution Selection

Contrary to conventional feature selection methods that yield a single solution,
MOEA produces an array of solutions. Consequently, selecting an suitable solution
is a crucial issue for MOEAs. Acknowledging that a minimal number of features
might lead to reduced accuracy during testing, PS-NSGA employs two additional
objectives, namely error rate and distance, to guide solution selection. Initially, a
subset of solutions demonstrating the lowest error rates on the training data is identified. From this subset, the solution exhibiting the minimal distance is chosen. More
discussions of solution selection is detailed in Sect. 3.4.3.

Table 3.1 Dataset

Dataset	#Features	#Instances	#Classes	% Smallest class	% Largest class
SRBCT	2,308	83	4	13	35
Leukemia 1	5,327	72	3	13	53
9Tumor	5,726	60	9	3	15
Brain Tumor 1	5,920	90	5	4	67
Prostate	10,509	102	2	49	51
Adenocarcinoma	9,868	76	2	16	84
Breast3	4,869	95	3	19	46
Nci	5,244	61	8	8	15
Prostate6033	6,033	102	2	49	51
Lymphoma	4,026	62	3	15	68

3.3   Experimental Design

3.3.1   Datasets

The efficacy of the newly developed PS-NSGA is assessed using ten benchmark gene datasets in our experimental setup. First five datasets are accessible via GitHub,[1] while the remaining five can be found in another repository.[2] Each dataset comprises samples containing gene microarray data along with the ground truth class label, denoting various disease types. The primary objective is to pinpoint the most pertinent genes for effective disease classification. Table 3.1 presents the details of the ten datasets, including the total number of features, samples, and classes, along with the percentage of samples in the largest and smallest classes. These datasets have a significantly higher number of features compared to the number of samples, and the gap between the largest and the smallest class is relatively large, so the datasets are highly unbalanced. For data visualization purposes, the t-distributed stochastic neighbor embedding algorithm(t-SNE) [23] is employed to condense the feature space. Figure 3.6 displays some of these visualizations,and the distribution of some datasets is chaotic, which adds to the challenge of classification and feature selection (FS).

---

[1] https://github.com/primekangkang/Genedata.

[2] https://github.com/rdiaz02/varSelRF-suppl-mat.

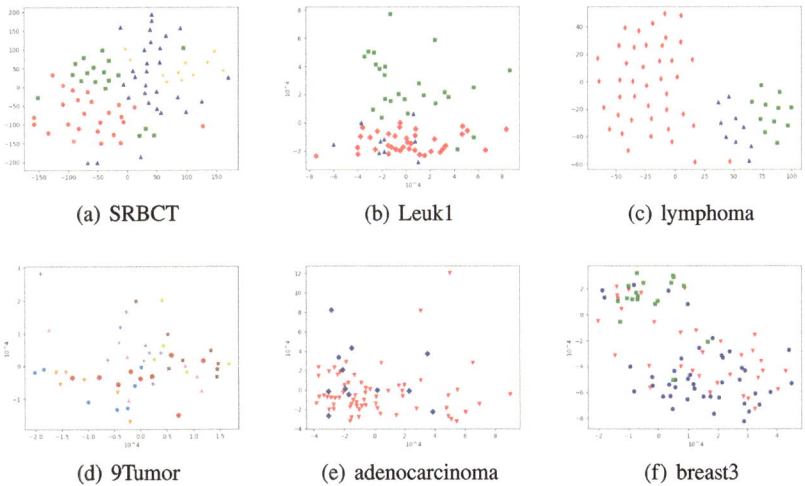

(a) SRBCT   (b) Leuk1   (c) lymphoma

(d) 9Tumor   (e) adenocarcinoma   (f) breast3

Fig. 3.6 Visualizaion of six datasets. The upper three datasets are relatively easy to separate different classes while the lower three are chaotic and hard to classify correctly

3.3.2 Experiment Settings

To ensure the correctness of our experiments, we employed a ten-fold cross-validation (CV) [24] to create training and testing datasets. In this method, each fold is sequentially utilized as a testing set and the remaining nine folds as the training set for Feature Selection (FS). Note that the testing set is not utilized during the FS. Once FS is complete, the optimal solution is chosen, following which both the training and testing sets are transformed by discarding the features that were not selected. And then we use the results on the transformed testing to evaluate algorithm. In FS process, we implement an an inner ten-fold cross-validation to calculate the individual's fitness values. Here, we use a K-Nearest Neighbors (KNN) classifier, setting K to 1, to measure the classification error rate. Given the significant imbalance present in our datasets, standard ten-fold cross-validation can lead to a skewed distribution of training data samples compared to the actual distribution, potentially hindering accurate predictions. Therefore, we use a stratified ten-fold cross validation to sample the data [24].

The experimental parameter configurations are detailed in Table 3.2. To ensure an equitable comparison, we have aligned the population size and the maximum number of iterations with those used in PPSO [25]. The other parameters are set when the best performances are obtained in applying GA in feature selection. As NSGAIII-FS operates under the same algorithmic framework as PS-NSGA, their parameter settings are identical. The settings for the other methods are suggested in the paper of PPSO and VLPSO [22, 25].

Table 3.2 Parameter Settings

Parameters	Settings
Population size	Features/20 (restriction to 300)
Number of mutation	Same as population size
Number of crossover	Same as population size
Maximum iteration	70
Divisions along each objective axis	27
Mutation probability	0.1
Number of mutation retry	1
Tolerance of accuracy	0.1

Table 3.1 reveals that the datasets contain between 2,000 and 11,000 features. Consequently, we have determined the population size to be 1/20 of the number of features, capped at a maximum of 300 to accommodate our computing resource constraints. Additionally, we have set the maximum number of iterations at 70. In PS-NSGA's association process, detailed in Fig. 3.5, every candidate is linked to a specific reference point. This association plays a crucial role in determining the candidate's likelihood of elimination. Consequently, it is essential for the distribution of reference points to be denser than that of the population to ensure effective decision-making. To accommodate this requirement, we set the number of reference points along each axis to 27. This decision allows us to produce 406 reference points, as calculated using Eq. 3.9. Additionally, the number of mutation operations and crossover operations is the same as the population size. We set the mutation rate to 0.1 and only allow one mutation retry. For each dataset, we performed 10 independent feature selections, each of which yielded an optimal solution. Since the outer ten-fold cross-validation produces 10 best solutions, we will get 100 solutions. We averaged the results of these solutions and compared them with other algorithms.

3.4 Results and Analysis

3.4.1 Comparisons of FS Results

To assess the effectiveness and performance of our newly developed PS-NSGA, we compare its classification capabilities with the original NSGA-III algorithm (referred to as NSGAIII-FS) and some state-of-the-art FS algorithms based on PSO. We also measure our method's performance against results obtained using the original feature set (denoted as $Full$). In NSGAIII-FS, a binary encoding system is utilized for chromosome representation and basic bit mutation is applied for mutation processes. PSO-FS follows a two-stage approach that first discretizes features using a Minimum Description Length (MDL) followed by standard PSO [26]. EPSO can simultaneously discretize and select features in a single stage and shows improvement in testing results [27]. PPSO, a recently introduced method, which applies a new representation to reduce the search space of the problem and a new fitness function to evaluate particles [25]. VLPSO, marking a first in PSO-based FS algorithms, uses variable-length representation and has shown superior performance compared to traditional fixed-length methods [22]. The results of the PSOs(PSO-FS, EPSO, PPSO, VLPSO) and *Full* for the first five datasets are derived from [22, 25]. We personally conducted the performance evaluation of NSGAIII-FS and MOEA/Dand the results of all methods in last five datasets (Adenocarcinoma, Breast3, Lymphoma, Nci, and Prostate6033) Additionally, we also compare our algorithm with another multi-objective algorithm based on MOEA/D [28]. Since MOEA/D is not directly applicable to our specific problem, we utilized a variant of MOEA/D, integrating PPSO as the search mechanism for FS, referred to as MOEA/D-FS.

All the results are presented in Table 3.3, with the smallest feature subset sizes and highest dataset accuracies highlighted in bold. It's important to note that, due to variations in algorithm frameworks, our parameter settings are unable to completely consistent with those of other algorithms, but the effectiveness of PSNSGA is still discernible as it shares the same settings for population size and maximum iterations as the compared methods. Furthermore, we have conducted a Wilcoxon significance test with 5% significance level to rigorously assess the performance of our proposed PS-NSGA against other techniques, where "+" indicates that our method significantly outperforms the compared method, "−" denotes inferior performance, and "=" signifies that the two methods yield similar results. PS-NSGA is the benchmark method in our analysis. The averaged results across these datasets are compiled and presented as *Overall* in Table 3.3.

From the data in Table 3.3, it's evident that PS-NSGA demonstrates a remarkable ability to reduce the number of features in all datasets. The algorithm selects feature subsets that are 2-3 orders of magnitude smaller than the complete feature sets. Not only does PS-NSGA select smaller subsets of features, but it also enhances accuracy. On average, across all datasets, PS-NSGA improves accuracy by 8.33%, while selecting merely 1.19% of the features.

In terms of feature subset size, PS-NSGA selects significantly smaller subsets compared to NSGAIII-FS, and in terms of test accuracy, PS-NSGA outperforms NSGAIII-FS in 9 out of 10 datasets, with the exception being Brain 1. and we also compare the efficiency of PS-NSGA and NSGAIII-FS, Fig. 3.7 presents the running times for executing the FS program on various datasets. The results show that PS-NSGA typically requires less time than NSGAIII-FS.

Compared with PSO-FS, PS-NSGA not only selects a smaller subset of features, but also achieves higher classification accuracy on all datasets. For example, in the prostate dataset, PS-NSGA selected only 65 features, while PSO-FS selected 777.4 features, while also outperforming PSO-FS by 4.24% in accuracy. Similarly, in the 9Tumor dataset, the subset of features selected by PS-NSGA was five times smaller than that of PSO-FS, and PS-NSGA had an accuracy rate of 11.96% higher than PSO-FS.

Table 3.3 Comparisons of FS results

Dataset		Full	NSGAIII-FS	PSO-FS	EPSO	PPSO	VLPSO	MOEA/D	PS-NSGA
SRBCT	Size:	2308	789.7	150.0	137.3	108.5	49.1	34.2	**18.6**
	Mean:	87.08	91.13	91.31	96.89	95.78	**99.67**	99.14	96.35
	S:	–	–	–	=	–	+	+	
Leuk1	Size:	5327	2085.2	150.0	135.9	80.4	54.7	32.1	**16.2**
	Mean:	79.72	84.44	81.60	93.37	94.37	93.31	**94.81**	91.98
	S:	–	–	–	+	+	+	+	
9Tumor	Size:	5726	2264.6	955.0	138.5	118.1	**44.2**	44.3	194.8
	Mean:	36.67	48.43	45.95	58.22	**59.28**	55.11	44.44	58.30
	S:	–	–	–	=	=	–	–	
Brain 1	Size:	5920	2335.2	317.3	150.7	73.4	**26.8**	83.1	57.8
	Mean:	72.08	**76.16**	71.00	72.79	74.40	71.19	74.27	73.81
	S:	=	=	–	=	=	–	=	
Prostate	Size:	10509	4556.1	777.4	54.9	65.6	**26.4**	55.9	65.0
	Mean:	85.33	84.72	85.20	83.74	**91.82**	89.82	90.46	89.44
	S:	–	–	–	–	+	=	+	
Adenocar-cinoma	Size:	9868	4172.3	236.5	89.6	18.1	67.5	**18.0**	59.7
	Mean:	62.74	60.00	63.35	66.73	57.90	60.12	57.73	**67.50**
	S:	–	–	–	=	–	–	–	
Breast 3	Size:	4869	2029.0	346.7	168.3	113.8	**32.9**	80.2	124.4
	Mean:	55.23	60.61	64.36	67.48	65.20	51.50	57.17	**68.02**
	S:	–	–	–	=	–	–	–	
Nci	Size:	5244	2012.1	255.4	189.2	133.5	121.3	117.0	**109.1**
	Mean:	68.26	71.69	70.26	75.37	77.85	77.38	63.81	**78.29**
	S:	–	–	–	–	=	–	–	
Prostate 6033	Size:	6033	2479.8	358.2	138.9	95.3	63.1	**49.0**	50.4
	Mean:	81.31	85.50	80.36	85.27	86.45	**89.16**	88.30	87.37
	S:	–	–	–	–	–	+	+	
Lymphoma	Size:	4026	1378.0	955	138.5	201.1	23.2	27.8	**18.7**
	Mean:	99.08	99.33	96.87	99.17	**100.00**	99.17	97.87	99.67
	S:	=	=	–	=	=	–	=	
Overall	Size:	5983	2410.2	450.2	134.2	100.8	**50.9**	54.2	71.4
	Mean:	72.74	76.20	66.50	79.90	80.30	78.64	76.80	**81.07**

In terms of dimensionality reduction, PS-NSGA selects a smaller features subject in 8 datasets compared to EPSO. In terms of classification accuracy, PS-NSGA surpassed EPSO in 8 datasets. It is worth noting that in the SRBCT dataset, although the difference in accuracy between the two algorithms is only 0.54%, PS-NSGA only selects selects 1/7 of features over EPSO.

PS-NSGA outperforms PPSO in accuracy across six datasets and selects notably smaller feature subsets in seven datasets. For instance, in the Leukemia 1 dataset,

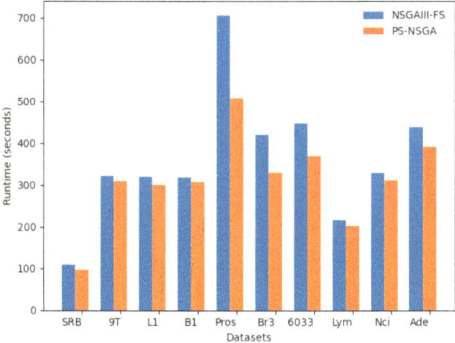

Fig. 3.7 Comparison of the running time in seconds

despite PPSO achieving 2.39% higher accuracy than PS-NSGA, the feature count selected by PS-NSGA is roughly one-fifth of that chosen by PPSO.

Compared with VLPSO, PS-NSGA selects fewer features than VLPSO in six datasets and achieved higher accuracy in six datasets. While VLPSO averages smaller feature subsets, PS-NSGA exceeds VLPSO in overall accuracy by 2.43%.

PS-NSGA and MOEA/D-FS exhibit strong performance on a variety of datasets. But overall, the average accuracy of PS-NSGA is 4.27% higher than that of MOEA/D-FS, and the number of features selected by PS-NSGA is 17.2 fewer than MOEA/D-FS.

In conclusion, PS-NSGA outperforms other multi-objective feature selection algorithms by achieving higher classification accuracy while selecting fewer features across multiple datasets. Notably, PS-NSGA consistently exhibits significantly higher accuracy in larger datasets and selects much fewer features in smaller datasets compared to other algorithms. When compared to several state-of-the-art PSO-based feature selection algorithms, PS-NSGA outperforms most algorithms in the size of feature subsets and achieves the highest average accuracy.

3.4.2 Comparisons with Traditional Methods

To evaluate whether PS-NSGA outperforms traditional feature selection methods, we compare it with three conventional approaches:linear forward selection (LFS), consistency-based search (CON) and correlation-based feature selection (CFS). Prior to feature selection, all three methods utilize MDL for feature discretization, where more detailed description about discretization and FS can be found in [29]. LFS, an enhancement of Sequential Forward Selection (SFS) and a wrapper method, reduces the number of features be added in each forward step [30]. CON, introduced in [31], is a consistency-based search approach utilized in Feature Selection (FS). On the other hand, CFS, as described in [32], is a correlation-based filter method that assesses

Table 3.4 Comparison results with traditional methods

Dataset		MDL+LFS	MDL+CON	MDL+CFS	PS-NSGA
SRBCT	Size:	6.1	**4.3**	80.9	18.6
	Mean:	88.75	85.83	**100**	96.35
	S:	–	–	+	
9Tumor	Size:	12.6	**7.6**	38	194.8
	Mean:	41.67	28.33	53.33	**58.3**
	S:	–	–	–	
Leukl	Size:	4.8	**3**	56	16.2
	Mean:	81.39	89.17	**93.19**	91.98
	S:	–	–	+	
Prostate	Size:	4.9	**4.7**	51.6	65
	Mean:	73.17	70.5	**90.17**	89.44
	S:	–	–	+	
Adenocarcinoma	Size:	4	**3.6**	37.4	59.7
	Mean:	59.03	46.68	64.87	**67.50**
	S:	–	–	–	
Breast3	Size:	7.5	**6.7**	47.6	124.4
	Mean:	57.45	33.76	**70.39**	68.02
	S:	–	–	+	
Nci	Size:	5.8	**5.6**	70.3	109.1
	Mean:	72.92	52.78	77.82	**78.29**
	S:	–	–	=	
Prostate	Size:	5.2	**4.8**	44.6	50.4
	Mean:	88.67	86.07	**89.78**	87.37
	S:	+	-	+	
Lymphoma	Size:	5.6	**4.8**	73.3	18.7
	Mean:	98.38	96.77	**100**	99.67
	S:	–	–	=	
Overall	Size:	5.7	**4.5**	50	70.9
	Mean:	66.14	59.0	**74.0**	73.6

and ranks subsets of features instead of individual features. The comparative results across all datasets are displayed in Table 3.4, highlighting the smallest feature subsets and highest accuracies in bold. Additionally, we apply a Wilcoxon significance test with 5% significance level on the results. The averaged data across these datasets are summarized as *Overall* in the Table 3.4.

According to Table 3.4, MDL+LFS selects a smaller number of features compared to PS-NSGA across all datasets; but its accuracy is lower than PS-NSGA in all datasets. This suggests that MDL+LFS tends to reach local optima prematurely, leading to smaller but less accurate feature subsets. Meanwhile, MDL+CON selects the fewest number of features overall, yet PS-NSGA outperforms it in accuracy across nine datasets. MDL+CFS surpassed PS-NSGA in terms of accuracy in six datasets, while PS-NSGA selected smaller feature subsets in three datasets.

In summary, PS-NSGA outperforms traditional feature selection methods across multiple datasets. When compared against traditional algorithms in 30 cases, PS-NSGA achieves higher average accuracy in 20 instances and selects fewer features in 16 instances. While PS-NSGA tends to select more features compared to MDL+LFS and MDL+CON, it surpasses these algorithms by over 6% in average accuracy. Although PS-NSGA lags behind MDL+CFS in both the number of selected features and average accuracy, the overall performance difference between the two methods is relatively small.

3.4.3 Studies on Solution Selection

As previously discussed, choosing an appropriate solution from the Pareto Front (PF) is crucial for Multi-Objective Evolutionary Algorithms (MOEAs). To assess the effect of the three objectives, we evaluated the testing accuracy of the PF after each run in the SRBCT dataset, which is the smallest dataset, making the impact more pronounced. We documented the training accuracy, testing accuracy, size of the feature subsets, and the distance of every member in PF.

To assess how training accuracy influences testing accuracy, we categorized the training accuracy into ten intervals, as illustrated in Fig. 3.8. We can see from the table that testing accuracy tends to rise with improved training accuracy. Nonetheless, situations often arise where several solutions attain maximum accuracy, indicating the necessity for an additional metric to identify the optimal single solution.

Figure 3.8 presents the testing accuracy for solutions varying in the size of their feature subsets. The goal of feature selection (FS) is to select the smallest feature subset while maintaining high performance. Yet, the data indicates that an overly small number of features might reduce performance. Consequently, the size of the feature subsets can not be the sole criterion in the selection of solutions.

Regarding the distance metric, the table clearly demonstrates that as the distance metric decreases, there is a continuous improvement in testing accuracy, indicating that the distance metric is a vital factor in selecting an appropriate solution.

From the discussion outlined earlier, we choose a subset of PF that achieves the lowest error rate in training data. Then, within this subset, the solution with the smallest distance is chosen as the final solution.In this strategy, the PF will be sorted twice by distance and training accuracy and the top-ranking solution is selected. To assess this method, we calculate the average performance across various intervals within the sorted PF, with the results depicted in Fig. 3.9.Generally, it's observed that the top solutions in the sorted PF are more likely to achieve better performance.

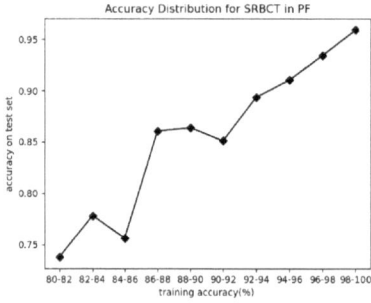

(a) The impact of the training accuracy

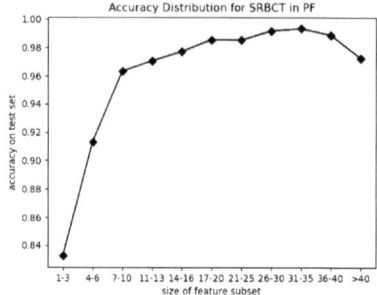

(b) The impact of the size of feature subsets

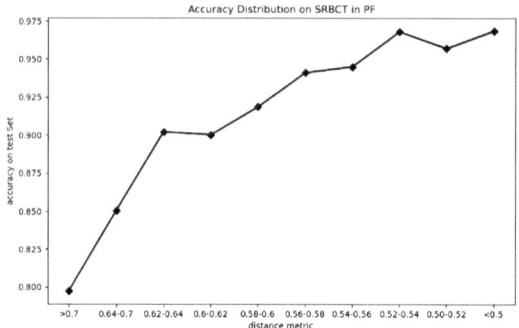

(c) The impact of the distance metric

Fig. 3.8 The impact of the training accuracy, size of feature subsets and distance metric on SRBCT

Fig. 3.9 Evaluation of the solution selection strategy on SRBCT

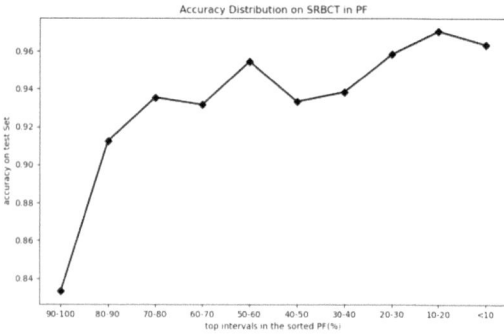

3.4.4 Impact of Accuracy-Preferred Domination

In PS-NSGA, we design an improved accuracy-preferred domination. To test its performance, we compared the average test results of the two domination methods, accuracy-preferred domination and basic domination, on all datasets after each iteration. The average accuracy and average number of features are shown in Fig. 3.10. In the experiments, all settings of the two algorithms are the same except for the domination method.

From Fig. 3.10, it's apparent that in the initial phase of the algorithm, due to the basic non-dominated sorting which does not favor any specific objective values, individuals with fewer features are easily retained, and this tendency to select fewer features leads to improved accuracy. Therefore, during this early stage, the average accuracy of the population increases at a similar rate for both methods.

In the later stages of the algorithm, having too few features often leads to a substantial decrease in accuracy. For the basic domination, individuals with few features but extremely low accuracy are not dominated by those with relatively more features

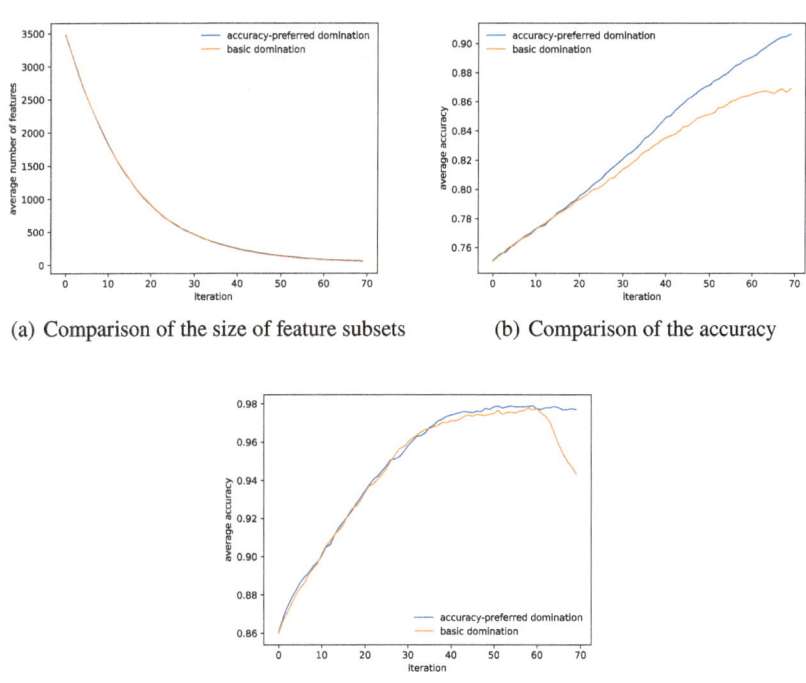

(a) Comparison of the size of feature subsets   (b) Comparison of the accuracy

(c) Comparison of the accuracy on SRBCT, which is typical for small datasets

Fig. 3.10 Comparison of accuracy-preferred domination and basic domination (average on all datasets)

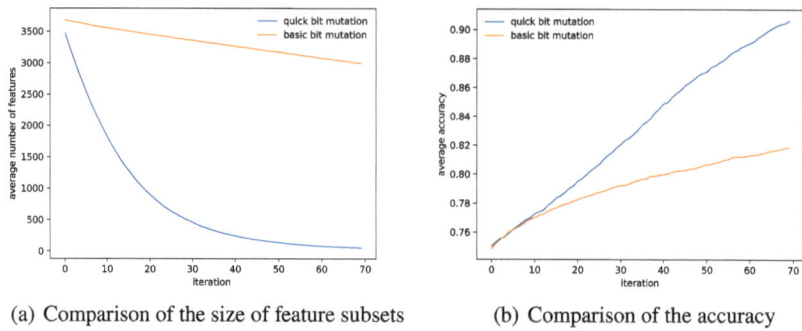

(a) Comparison of the size of feature subsets   (b) Comparison of the accuracy

Fig. 3.11 Comparison of quick bit mutation and basic bit mutation (average on all datasets)

but high accuracy, leading to a reduction in average accuracy. For the accuracy-preferred domination, individuals with lower accuracy are dominated by those with higher accuracy, so they are more likely to be eliminated, which will result in the lead of accuracy-preferred domination in the later stage of algorithm.The results indicate that accuracy-preferred domination enhances the performance of the algorithm and the diversity of the population during the evolutionary process. Additionally, it mitigates the risk of the algorithm becoming trapped in local optima, thereby ensuring a more robust and effective feature selection.

3.4.5   Impact of Quick Bit Mutation

We also designed comparative experiments of quick bit mutation and basic bit mutation to compare their performance. The results are shown in Fig. 3.11.

   The figure shows that basic bit mutation reduce the number of features slowly. This is attributed to the fact that the positions of its mutations are chosen completely randomly, so that the number of 0s and 1s in the chosen positions is almost equal. Therefore, after mutation, the balance between the quantities of 0s and 1s remains, leading to inefficient mutation and slow accuracy growth. In contrast, quick bit mutation overcomes the limitation of probability and improves the efficiency of mutation. This results in a faster improvement in accuracy.

References

1. Zhou, Y., Zhang, W., Kang, J., Zhang, X., Wang, X.: A problem-specific non-dominated sorting genetic algorithm for supervised feature selection. Inf. Sci. **547**, 841–859 (2021)
2. James, G., Witten, D., Hastie, T., Tibshirani, R.: An Introduction to Statistical Learning, vol. 112. Springer (2013)

3. John, G.H., Kohavi, R., Pfleger, K.: Irrelevant features and the subset selection problem. In: Machine Learning Proceedings 1994, pp. 121–129 (1994)
4. Whitney, A.W.: A direct method of nonparametric measurement selection. IEEE Trans. Comput. **100**(9), 1100–1103 (1971)
5. Marill, T., Green, D.: On the effectiveness of receptors in recognition systems. IEEE Trans. Inf. Theory **9**(1), 11–17 (1963)
6. Kudo, M., Sklansky, J.: Comparison of algorithms that select features for pattern classifiers. Pattern Recogn. **33**(1), 25–41 (2000)
7. Dong, H., Li, T., Ding, R., Sun, J.: A novel hybrid genetic algorithm with granular information for feature selection and optimization. Appl. Soft Comput. **65**, 33–46 (2018)
8. Zhang, Y., Gong, D., Cheng, J.: Multi-objective particle swarm optimization approach for cost-based feature selection in classification. IEEE/ACM Trans. Comput. Biol. Bioinform. (TCBB) **14**(1), 64–75 (2017). (IEEE Computer Society Press)
9. Tran, B., Zhang, M., Xue, B.: A PSO based hybrid feature selection algorithm for high-dimensional classification. In: 2016 IEEE Congress on Evolutionary Computation (CEC), pp. 3801–3808. IEEE (2016)
10. Khushaba, R.N., Al-Ani, A., Al-Jumaily, A.: Feature subset selection using differential evolution and a statistical repair mechanism. Expert Syst. Appl. **38**(9), 11515–11526 (2011)
11. Taradeh, M., Mafarja, M., Heidari, A.A., Faris, H., Aljarah, I., Mirjalili, S., Fujita, H.: An evolutionary gravitational search-based feature selection. Inf. Sci. **497**, 219–239 (2019). https://doi.org/10.1016/j.ins.2019.05.038
12. Zhang, Y., Song, X., Gong, D.: A return-cost-based binary firefly algorithm for feature selection. Inf. Sci. **418–419**, 561–574 (2017). https://doi.org/10.1016/j.ins.2017.08.047
13. Al-Sahaf, H., Zhang, M., Johnston, M., Verma, B.: Image descriptor: a genetic programming approach to multiclass texture classification. In: 2015 IEEE Congress on Evolutionary Computation (CEC), pp. 2460–2467 (2015)
14. Zaretalab, A., Hajipour, V., Sharifi, M., Shahriari, M.R.: A knowledge-based archive multi-objective simulated annealing algorithm to optimize series–parallel system with choice of redundancy strategies. Comput. Ind. Eng. **80**, 33–44 (2015)
15. Zhu, Y., Liang, J., Chen, J., Ming, Z.: An improved NSGA-III algorithm for feature selection used in intrusion detection. Knowl. Based Syst. **116**, 74–85 (2017). (Elsevier)
16. Hancer, E., Xue, B., Zhang, M., Karaboga, D., Akay, B.: Pareto front feature selection based on artificial bee colony optimization. Inf. Sci. **422**, 462–479 (2018). https://doi.org/10.1016/j.ins.2017.09.028
17. Deb, K., Jain, H.: An evolutionary many-objective optimization algorithm using reference-point-based nondominated sorting approach, part I: solving problems with box constraints. IEEE Trans. Evol. Comput. **18**(4), 577–601 (2013)
18. Ding, R., Dong, H., He, J., Feng, X., Yu, X., Li, L.: U-NSGA-III: an improved evolutionary many-objective optimization algorithm. In: International Conference on Bio-Inspired Computing: Theories and Applications, pp. 24–35 (2018)
19. Yi, J.-H., Deb, S., Dong, J., Alavi, A.H., Wang, G.-G.: An improved NSGA-III algorithm with adaptive mutation operator for big data optimization problems. Futur. Generat. Comput. Syst. **88**, 571–585 (2018)
20. Masood, A., Chen, G., Mei, Y., Zhang, M.: Adaptive reference point generation for many-objective optimization using NSGA-III. In: Australasian Joint Conference on Artificial Intelligence, pp. 358–370 (2018)
21. Vesikar, Y., Deb, K., Blank, J.: Reference point based NSGA-III for preferred solutions. In: 2018 IEEE Symposium Series on Computational Intelligence (SSCI), pp. 1587–1594 (2018)
22. Tran, B., Xue, B., Zhang, M.: Variable-length particle swarm optimization for feature selection on high-dimensional classification. IEEE Trans. Evol. Comput. **23**(3), 473–487 (2019)
23. van der Maaten, L., Hinton, G.: Visualizing data using t-SNE. J. Mach. Learn. Res. **9**(Nov), 2579–2605 (2008)
24. Kohavi, R., et al.: A study of cross-validation and bootstrap for accuracy estimation and model selection. In: Ijcai, vol. 14, no. 2, pp. 1137–1145. Montreal, Canada (1995)

25. Tran, B., Xue, B., Zhang, M.: A new representation in PSO for discretization-based feature selection. IEEE Trans. Cybern. **48**(6), 1733–1746 (2017). (IEEE)
26. Kennedy, J., Eberhart, R.: Particle swarm optimization. In: Proceedings of ICNN'95-International Conference on Neural Networks, vol. 4, pp. 1942–1948. IEEE (1995)
27. Tran, B., Xue, B., Zhang, M.: Bare-bone particle swarm optimisation for simultaneously discretising and selecting features for high-dimensional classification. In: European Conference on the Applications of Evolutionary Computation, pp. 701–718. Springer (2016)
28. Zhang, Q., Li, H.: MOEA/D: a multiobjective evolutionary algorithm based on decomposition. IEEE Trans. Evol. Comput. **11**(6), 712–731 (2007)
29. Tsai, C.-F., Chen, Y.-C.: The optimal combination of feature selection and data discretization: an empirical study. Inf. Sci. **505**, 282–293 (2019). https://doi.org/10.1016/j.ins.2019.07.091
30. Gutlein, M., Frank, E., Hall, M., Karwath, A.: Large-scale attribute selection using wrappers. In: 2009 IEEE Symposium on Computational Intelligence and Data Mining, pp. 332–339. IEEE (2009)
31. Dash, M., Liu, H.: Consistency-based search in feature selection. Artif. Intell. **151**(1–2), 155–176 (2003). (Elsevier)
32. Hall, M.A.: Correlation-based feature selection of discrete and numeric class machine learning. University of Waikato, Department of Computer Science (2000)

Chapter 4
Evolutionary Algorithm Based Local Feature Selection

Abstract This chapter introduces the concept of local feature selection (LFS) as a contrast to traditional methods. Unlike existing LFS algorithms, which use distance-like objective functions, we propose a region purity-based LFS (RP-LFS) (Zhou et al. in IEEE Trans Evol Comput (2023), [1]) that incorporates a novel objective function, region purity, for multi-objective optimization. RP-LFS partitions the sample space into local regions and obtains a feature subset for each region, resulting in improved classification accuracy. To solve the RP-LFS problem, we propose an improved non-dominated sorting genetic algorithm III, which includes a network-inspired crossover operator and a quick bit mutation. We also develop a regional feature sharing strategy to preserve effective features between different local models. Experimental studies on 11 UCI datasets and nine high-dimensional datasets confirm the effectiveness of RP-LFS. Compared to state-of-the-art feature selection and LFS algorithms, RP-LFS achieves competitive classification accuracy while reducing the feature subset size.

Keywords Local feature selection · Region purity · Non-dominated sorting genetic algorithm III

4.1 Problem Formulation

Many machine-learning applications have faced challenges due to the increase in high-dimensional data, which often include excessive redundant and irrelevant features that can impair algorithm performance and increase computation time. Thus, feature selection (FS) is a crucial preprocessing step that aims to enhance classification performance by selecting the optimal subset of features from a pool of redundant ones, and is of significant importance.

A FS algorithm usually consists of two main components: the search for good feature subsets and the evaluation of these subsets [2]. Different methodologies, such as heuristic search algorithms [3–5], have been used to search for feature subsets. However, due to the large search space, especially for high-dimensional datasets, the search process can easily get stuck in a local optimum. To address this challenge, evolutionary algorithms (EAs) like particle swarm optimization (PSO) [6],

© The Author(s), under exclusive license to Springer Nature Singapore Pte Ltd. 2025
Y. Zhou et al., *Computational Intelligence for High-Dimensional Machine Learning*,
SpringerBriefs in Computer Science, https://doi.org/10.1007/978-981-96-2687-8_4

genetic programming (GP) [7], differential evolution (DE) [8], and genetic algorithm (GA) [9] are utilized for their powerful search capabilities and better convergence. On the other hand, feature subset evaluation aims to create an evaluation criterion to determine the quality of the chosen feature subset. In supervised learning, training accuracy in terms of classification error is a common criterion. Other criteria, such as feature subset size [10], redundancy [11], relevance [12], or distance metric [13], are combined with training error to create a more effective objective for optimization. Traditional approaches often use a weighted sum to aggregate objectives, which requires determining and adjusting the weights. To overcome this issue, directly optimizing multiple objectives [14–16] in FS has been explored and found to be more efficient in subset selection and classification performance.

Traditional feature selection method assumes that a subset of features can be applied to all samples, resulting in a global subset of features. However, in practical applications, selected features may only be relevant in specific local regions, and using a global feature subset may not be effective when data distributions vary. To address this, local feature selection (LFS) has been proposed [17–20]. LFS partitions training data into different local regions and selects the best feature subsets for each region. Unlike global feature selection, LFS generates diverse feature subsets based on different local regions. However, directly using training accuracy as an evaluation metric can introduce bias and unfairness. Thus, finding a reliable local feature evaluation metric is challenging. Recent studies have optimized LFS by maximizing the distance between different classes while minimizing the distance [18–20] within each class for all training samples. The classification result of LFS is determined by the similarity between a testing sample and training samples from different classes. However, relying solely on distance is not sufficient for selecting the most effective features for classification, as illustrated and explained in Fig. 4.1.

In the evaluation phase, we examine a binary class problem with a feature set $\{f_1, f_2, f_3, f_4\}$ and analyze the distribution of data samples, as shown in Fig. 4.1. Figure 4.1a illustrates the distribution of data on the feature subset $\{f_1, f_2\}$, where the samples are divided into three distinct local regions based on their distances. Each local region is determined by a sample point with a predefined radius, and for simplicity, we depict them as non-overlapping divisions. Consider local region 1 as an example. In the testing phase, if a query sample, indicated by the blue arrow, falls into local region 1 and its nearest neighboring sample is labeled as Class 1, it will be misclassified as Class 1. This misclassification occurs because the query sample is assigned based on its proximity to the nearest neighbor. Figure 4.1b presents the distribution of samples from local region 1 projected onto the feature subset $\{f_3, f_4\}$. Although the sample distribution is sparse, no misclassification occurs because a higher concentration of samples belonging to the same class ensures correct classification. However, this issue becomes more significant in higher-dimensional spaces. Hence, a critical challenge lies in developing a new criterion that can effectively aid feature selection within local regions, ultimately leading to improved classification accuracy.

Recently, there have been advancements in the field of feature selection, specifically using multiobjective evolutionary algorithms (MOEAs) [21, 22] to solve the

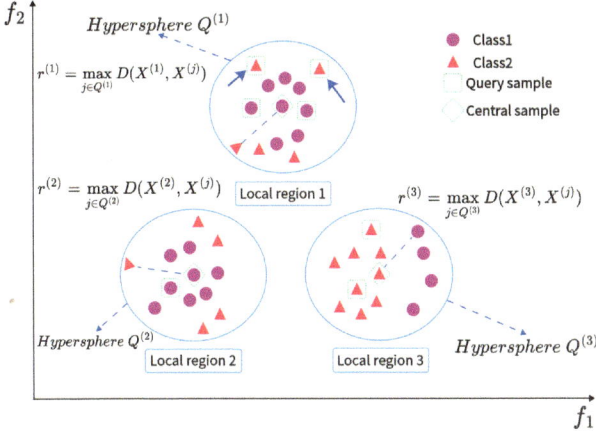

(a) Data sample distribution projected on feature subset $\{f_1, f_2\}$.

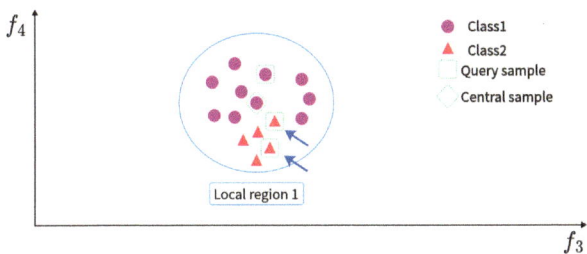

(b) Data sample distribution of local region 1 projected on feature subset $\{f_3, f_4\}$.

Fig. 4.1 A toy example of LFS for a binary class problem

problem as multi-objective optimization problems (MOPs) [8, 10, 16, 23]. This approach involves optimizing two or more conflicting objectives simultaneously. By incorporating multiple objectives, we can gain a better understanding of the essential properties of feature selection problems [24, 25] and ultimately improve the classification performance. One popular MOEA, known as NSGA-III, has shown great potential in solving MOPs with three or more objectives [22]. It combines the advantages of a non-dominated sorting framework and a decomposition-based method, resulting in both good convergence and diversity. NSGA-III and its variants [26, 27] have been successfully applied in various domains, including engineering [28], machine learning [29], and operational research [30], demonstrating their effectiveness.

In this chapter, we propose RP-LFS (Region Purity-based Local Feature Selection), which involves three key objectives. The first objective is to design a novel criterion called RP (Region Purity), which aims to retain as many samples from the same class as possible in the local region, while excluding samples from different classes that may lead to misclassification. We also consider a region-based distance metric to enhance generalization. The second objective is to dynamically adjust the number of selected features using a feature subset size criterion, ensuring better adaptability. To address the formulated MOP (Multi-Objective Problem), an advanced MOEA (Multi-Objective Evolutionary Algorithm) specifically designed for LFS is proposed within the framework of NSGA-III.To enhance population diversity, we introduce a network-inspired crossover operator that addresses the limitations of fully random selection of parents. Additionally, we develop a quick bit mutation operator that efficiently converges when dealing with high-dimensional features. To expedite convergence within each local region, we propose a regional feature sharing strategy (RFSS) that leverages feature push and pop operators. These operators dynamically update the local region based on FS (Feature Selection) outcomes. Finally, in order to ensure improved test accuracy, both RP and the region-based distance metric are considered for generality.Experimental studies conducted on 11 UCI datasets and nine high-dimensional datasets demonstrate the superiority of our proposed RP-LFS over state-of-the-art LFS methods, as well as traditional and evolutionary FS algorithms, in terms of classification accuracy and feature subset size. In summary, the contributions to this chapter are as follows:

- To the best of our knowledge, this is the first work to explore the use of a regional objective function in local feature selection (LFS). We introduce RP as a solution to address the limitation of relying solely on distance-based criteria.
- To improve the selection of local feature subsets, we have created an evolutionary multi-objective optimization framework named RP-LFS. This framework considers three objectives: RP, a distance metric based on region, and the proportion of selected features.
- To address the formulated multi-Objective problem (MOP) in each local region, we introduce a novel multi-Objective evolutionary algorithm (MOEA) based on the NSGA-III framework. This algorithm incorporates a network-inspired crossover operator, a quick bit mutation operator, and a regional feature sharing strategy among different local regions.

4.2 Local Region Partition

An overview of the RP-LFS algorithm is shown in Fig. 4.2. In this algorithm, local zoning is performed first, i.e., each sample in the training set is taken as the central sample and assigned to a local region where all samples share the same selected feature. Then, multi-objective optimization was performed using the improved NSGA-III for each localized region to obtain a subset of features for that region. Next, the regional feature sharing policy is enforced, updating all feature subsets until the stop condition is met.

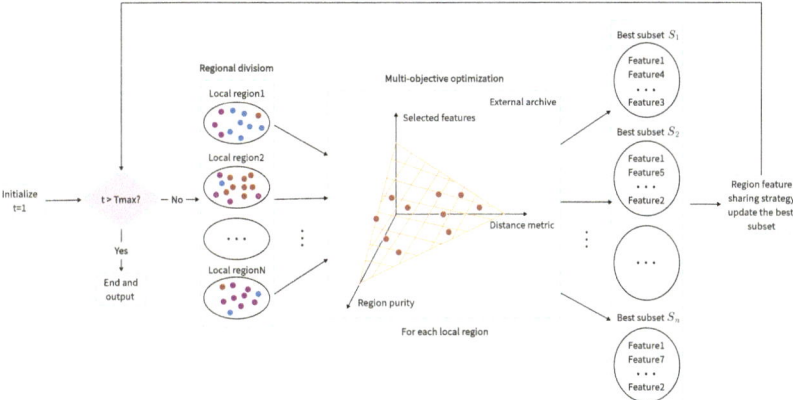

Fig. 4.2 Overview of our proposed method

Unlike existing LFS methods, in RP-LFS, zoning is performed first, followed by an LFS assessment. Specifically, in RP-LFS, we perform a locoregional design before LFS. We divide the entire region into N local regions, where N represents the number of training samples. We then treat each sample as a central sample and construct the corresponding hypersphere as a local region i with a radius based on impurities. The process consists of two main steps: first, the sample represented by the supersphere $Q^{(i)}$ in the local region a is determined according to Eq. (4.1):

$$\begin{cases} min & \delta(Q^{(i)}) \\ s.t. & \upsilon(Q^{(i)}) \geq \gamma \end{cases} \tag{4.1}$$

where $\delta(\cdot)$ represents the number of samples within the hypersphere, $\upsilon(\cdot)$ is the metric used to calculate the impurity. Impurity is defined as the ratio of the normalized number of samples with a different class label to the number of samples with the same label within the hypersphere. Additionally, we introduce a constant γ to adjust the level of impurities. Furthermore, we establish the maximum distance between the central sample and the other samples within the hypersphere to be equal to the radius of the local region i, as shown in Eq. (4.2):

$$r^{(i)} = max_{j \in Q^{(i)}, j \neq i}(D(x^{(i)}, x^{(j)})) \tag{4.2}$$

where $r^{(i)}$ represents the radius of local region i, $x^{(i)}$ denotes the center sample in local region i, i represents the index of the sample within the hypersphere $Q^{(i)}$, and $D(x^{(i)}, x^{(j)})$ denotes the Euclidean distance between two samples $x^{(i)}$ and $x^{(j)}$. To solve Eqs. (4.1) and (4.2), we assume that we have a central training sample $x^{(i)}$ with a class label $y^{(i)}$ and a corresponding feature set $f^{(i)}$. We divide all of the training samples into two groups based on their label, the ones with the same label as the central sample (denoted by purple circle) and those with a different label

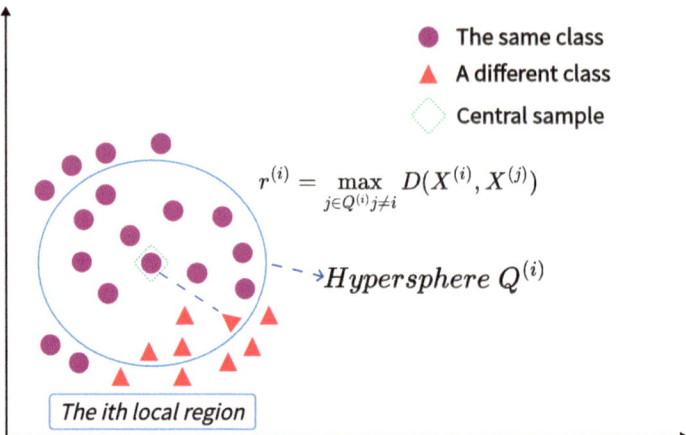

Fig. 4.3 Determination of the local region in RP-LFS

(denoted by red triangle). In the feature set $f^{(i)}$, we calculate the distance between the central sample and the rest of the samples. Then, we select samples whose distance is close to the central sample and gradually calculate the impurity υ based on their corresponding labels until $\upsilon \geq \gamma$. Equation (4.2) represents the distance between the last sample that satisfies Eq. (4.1) and the central sample, which is defined as the radius. Subsequently, we define a hypersphere $Q^{(i)}$ as the local region i in the co-ordinate system defined by $f^{(i)}$, centered at $x^{(i)}$, where the samples within the local region are determined by Eq. (4.1) and the radius is determined by Eq. (4.2), as shown in Fig. 4.3.

4.3   LFS Based on MOEA

4.3.1   Encoding and Update Strategy

4.3.1.1   Fitness Function

In contrast to conventional GFS methods, LFS involves the presence of multiple local feature subsets, thereby necessitating the adoption of a specialized aggregating classification model instead of traditional classifiers. However, this leads to fairness issues and significantly increases the time required for solution evaluation in genetic algorithms (GAs). Consequently, employing classification accuracy as a criterion during the training process becomes impractical. To enhance the performance of LFS, RP-LFS addresses the following three objectives, which are to be minimized.

(1) Region purity:

In order to maintain a high number of samples from the same class within a local region, while excluding samples from a different class that may lead to misclassification, we introduce a novel objective function called RP. Once the local region i is determined, we define the relevant indicators for the samples within this region, as described by Eqs. (4.3) and (4.4).

$$sc^{(i)} = \sum_{j \in Q^{(i)}} \zeta_i \left(\text{K-Neighbour}(\mathbf{x}^{(j)}), \mathbf{x}^{(i)} \right) \tag{4.3}$$

where $\text{K-Neighbour}(\mathbf{x}^{(j)})$ represents $\mathbf{x}^{(j)}$ along with its K-nearest samples within $Q^{(i)}$, and $\zeta(\cdot)$ serves as an indicator function that verifies whether $\mathbf{x}^{(j)}$ and its nearest K samples possess the same label as $\mathbf{x}^{(i)}$. If they do, $\zeta(\cdot)$ is assigned a value of 1; otherwise, it is set to 0. In the case when $\zeta(\cdot) = 1$, $\mathbf{x}^{(j)}$ is considered a positive sample. As a result, $sc^{(i)}$ denotes the count of positive samples within $Q^{(i)}$.

$$dc^{(i)} = \sum_{j \in Q^{(i)}} \eta_i \left(\text{K-Neighbour}(\mathbf{x}^{(j)}), \mathbf{x}^{(i)} \right) \tag{4.4}$$

where $\eta(\cdot)$ is an indicator that determines whether $\mathbf{x}^{(j)}$ and its nearest K samples have a different label from that of $\mathbf{x}^{(i)}$. If $\eta(\cdot) = 1$, it means that $\mathbf{x}^{(j)}$ is considered a negative sample. Conversely, if $\eta(\cdot) = 0$, it means that $\mathbf{x}^{(j)}$ has the same label as $\mathbf{x}^{(i)}$. Therefore, $dc^{(i)}$ represents the number of negative samples within the local region $Q^{(i)}$. To address the issue of imbalanced samples, $sc^{(i)}$ and $dc^{(i)}$ are finally normalized, as described below:

$$rp^{(i)} = \frac{1 + \dfrac{dc^{(i)}}{\sum_{j=1}^{N} \sigma(y^{(j)} \neq y^{(i)})} - \dfrac{sc^{(i)}}{\sum_{j=1}^{N} \sigma(y^{(j)} = y^{(i)})}}{2} \tag{4.5}$$

where $\sigma(\cdot)$ is an indicator function, representing whether a certain condition is satisfied. If the condition holds true, then $\sigma(\cdot) = 1$; otherwise, $\sigma(\cdot) = 0$. It is important to note that N denotes the total number of training samples. The variable $rp^{(i)}$, which is a value between 0 and 1, is subject to certain constraints. Specifically, when all the samples within $Q^{(i)}$ that possess the label $y^{(i)}$ are considered positive samples, and the remaining samples with differing labels are not designated as negative samples, the value of $rp^{(i)}$ is set to 0. On the other hand, when all the samples within $Q^{(i)}$ that possess a label different from $y^{(i)}$ are regarded as negative samples, and the samples belonging to $y^{(i)}$ in $Q^{(i)}$ are not identified as positive samples, $rp^{(i)}$ takes a value of 1. In theory, a higher value of $sc^{(i)}$ and a lower value of $dc^{(i)}$ indicate a greater purity among the samples within $Q^{(i)}$, implying the presence of a common feature subset within the local region. It should be emphasized that in order to align with the other two objectives that require minimization, we aim to minimize Eq. (4.5) to properly reflect the RP metric.

(2) Proportion of the selected features:

To enhance the efficacy of dimensionality reduction, we take into account the proportion of selected features, denoted as *psf*. This parameter is defined as follows:

$$psf^{(i)} = \frac{L^{(i)}}{D} \tag{4.6}$$

where $L^{(i)}$ denotes the number of chosen features within the local region indexed by i, and D denotes the overall number of features that are available.

(3) Region-based distance metric:

The objective of employing the region-based distance metric (denoted by *rdm*) is to investigate the delineation between samples belonging to distinct classes, ultimately enhancing generalization within each local region $Q^{(i)}$. The metric is defined as follows:

$$D_b^{(i)} = \min_{\{j|j\in Q^{(i)}, j\neq i, y^{(i)}\neq y^{(j)}\}} \mathrm{Dis}\left(x^{(i)}, x^{(j)}\right) \tag{4.7}$$

$$D_w^{(i)} = \max_{\{j|j\in Q^{(i)}, j\neq i, y^{(i)}=y^{(j)}\}} \mathrm{Dis}\left(x^{(i)}, x^{(j)}\right) \tag{4.8}$$

$$rdm^{(i)} = \frac{1}{1 + \exp^{-5\left(D_w^{(i)} - D_b^{(i)}\right)}} \tag{4.9}$$

where j represents the index of the sample within $Q^{(i)}$, $Dis(x^{(i)}, x^{(j)})$ denotes the Manhattan distance between two samples $x^{(i)}$ and $x^{(j)}$, D_b indicates the distance between the central sample and the closest sample with a different class, and D_w represents the distance between the central sample and the farthest sample within the same class. The value -5 in Eq. (4.9) is utilized to restrict the effective range of the distance difference in the sigmoid function, as recommended in [31].

4.3.1.2 Chromosome Encoding

Our approach utilizes binary bit strings to represent chromosomes, where each feature is assigned either 1 to indicate selection or 0 otherwise. Considering a dataset possessing D features, the ith chromosome is encoded as a D-dimensional string within the population, as depicted below:

$$x^{(i)}(t) = \left(x_1^{(i)}, x_2^{(i)}, \dots, x_j^{(i)}\right)$$
$$x_j^{(i)} \in \{0, 1\}, j = 1, 2, \dots, D, i = 1, 2, \dots, N \tag{4.10}$$

where $x^{(i)}(t)$ represents the ith chromosome in the tth generation, N denotes the population size, and $x_j^{(i)} \in \{0, 1\}$ denotes the binary encoding of the jth gene in the ith chromosome.

4.3.2 Feature Sharing Mechanism

Several correlation-guided methods [32, 33] have demonstrated their effectiveness in aiding the evolutionary process. As each local region is defined by a hypersphere anchored to a central training sample, it is possible for these regions to overlap. It is logical to assume that these overlapped regions frequently exhibit shared features. To harness this characteristic, we propose a regional feature sharing strategy during the search for local feature subsets. The goal of this strategy is to leverage the features selected by the overlapping region to enhance the performance of its respective local region and therefore achieve improved solutions.

Prior to outlining the strategy for regional feature sharing, it would be beneficial to provide clear definitions for key concepts related to each specific local region. Taking local region i as an illustration, the overlap region i can be defined as the collection of samples belonging to local region i that share the same label as the ith central sample. Consequently, every sample within overlap region i possesses its respective local feature subset. Let us assume that the feature set, denoted as $OR^{(i)}$, represents the collective set of these local feature subsets within the overlap region. Additionally, let $Uf^{(i)}$ denote the compilation of all features selected by overlap region i, excluding those associated with the local region i. It can thus be observed that $Uf^{(i)} = OR^{(i)} \setminus f^{(i)}$.

4.3.2.1 Feature Push Operator

In the feature push operator, the contribution of a feature to its corresponding local region is determined by its frequency of occurrence in the overlapping region. To accomplish this, the features in $Uf^{(i)}$, which belong to the local region i, are sorted in descending order based on their occurrence frequency in the local feature subsets of the overlap region i. Subsequently, each feature in the sorted $Uf^{(i)}$ is systematically added to $f^{(i)}$ to create a new local feature subset $f_{temp}^{(i)}$. The feature push operator terminates and $f_{temp}^{(i)}$ replaces $f^{(i)}$ when the $rp^{(i)}$ of the local region i using $f_{temp}^{(i)}$ is lower than that of $f^{(i)}$, or when $rp^{(i)}$ is equal but $f_{temp}^{(i)}$ has a smaller region-based distance metric. It is important to note that if no update of $f^{(i)}$ occurs after attempting all the features in $Uf^{(i)}$, the feature push operator ceases, leaving $f^{(i)}$ unchanged.

Take three samples in local region i as an example, assuming they have the same class label as the central sample. The local feature subset obtained is $f^{(i)} = \{f_1, f_4\}$. The local feature subset of the three samples in the overlap region i are $f^{(1)} = \{f_1, f_3\}$, $f^{(2)} = \{f_2, f_3, f_4\}$, and $f^{(3)} = \{f_2, f_3, f_5\}$ when they are

Input: The size of training set N, current best feature subset for the local region i,
$\quad\quad f^{(i)}, i = 1, ..., N$, the sample set in local region i, $R^{(i)}, i = 1, ..., N$
Output: Update the selected feature subset for the local region i, $f^{(i)}, i = 1, ..., N$
while $i < N$ **do**
$\quad$ **for** each sample from $R^{(i)}$ **do**
$\quad\quad |\quad OR^{(i)}$ = Compute all selected feature in the overlap region i.
$\quad$ **end**
$\quad Uf^{(i)}$ = Exclude all the features of $f^{(i)}$ within $OR^{(i)}$
$\quad$ Descending order for $Uf^{(i)}$ according the frequency of selected feature in overlap
$\quad\quad$ region i.
$\quad$ **for** Select k from $Uf^{(i)}$ **do**
$\quad\quad f^{(i)}_{temp}$ = add k into $f^{(i)}$
$\quad\quad$ Compute the fitness of $f^{(i)}_{temp}$
$\quad\quad$ Compare the fitness of $f^{(i)}$ and $f^{(i)}_{temp}$
$\quad\quad$ **if** meet the condition **then**
$\quad\quad\quad | \quad f^{(i)} = f^{(i)}_{temp}$
$\quad\quad\quad$ break
$\quad\quad$ **end**
$\quad$ **end**
end

Algorithm 1: Feature push operator

regarded as the central sample. Therefore, the feature set of the overlap region $OR^{(i)} = \{f_1, f_2, f_3, f_4, f_5\}$, and the set of unique features $US_i = \{f_2, f_3, f_5\}$. Sorting these features in $Uf^{(i)}$ in descending order based on their appearing frequency, we have $\{f_3, f_2, f_5\}$. We start by adding f_3 to form $\{f_1, f_3, f_4\}$. If $rp^{(i)}(\{f_1, f_3, f_4\}) < rp^{(i)}(\{f_1, f_4\})$, we stop the feature push operator and update the local feature subset as $f^{(i)} = \{f_1, f_3, f_4\}$, where $rp^{(i)}(\cdot)$ represents the function to calculate the RP (Region Proposal) of local region i with a certain feature subset. Otherwise, we continue by adding f_2 to form $\{f_1, f_2, f_4\}$ for evaluation, and so on. We iterate through the features one by one until the stopping criterion is met. The pseudocode for the feature push operator is presented in Algorithm 1.

4.3.2.2 Feature Pop Operator

In the feature pop operator, we aim to identify features that are less likely to be beneficial to the local region based on their frequency of occurrence in the corresponding overlap region. To achieve this, we sort the features in the local feature subset $f^{(i)}$ in ascending order, according to the number of times they appear in the local feature subsets of the overlap region i. Next, we iteratively remove each feature from the sorted $f^{(i)}$ to create a temporary feature set $f^{(i)}_{temp}$. After each removal, we evaluate the performance of the local region i by calculating its $rp^{(i)}$ using $f^{(i)}_{temp}$. If the $rp^{(i)}$ with $f^{(i)}_{temp}$ is lower than that with the original $f^{(i)}$, or if it is the same but has a smaller region-based distance metric, we stop the feature pop operator. In this case, $f^{(i)}_{temp}$

replaces $f^{(i)}$ as the new local feature subset. It is important to note that in the event that there is no update to $f^{(i)}$ after considering all the features in $f^{(i)}$, the feature push operator is also halted, and $f^{(i)}$ remains unchanged.

Input: Size of the training set N, current best feature subset for the local region i,
        $f^{(i)}, i = 1, ..., N$, the sample set in local region i, $R^{(i)}, i = 1, ..., N$
Output: Updated the selected feature subset for the local region i, $f^{(i)}, i = 1, ..., N$
while $i < N$ **do**
    for each sample from $R^{(i)}$ **do**
        | $OR^{(i)}$ = Compute all selected feature in the overlap region i.
    end
    Ascending order for $f^{(i)}$ according the frequency of selected feature in the overlap
    region i.
    for Select k from $f^{(i)}$ **do**
        $f_{temp}^{(i)}$ = remove k from $f^{(i)}$
        Compute the fitness of $f_{temp}^{(i)}$
        Compare the fitness of $f^{(i)}$ and the fitness of $f_{temp}^{(i)}$
        if meet the conditions **then**
        | $f^{(i)} = f_{temp}^{(i)}$
        | break
        end
    end
end

Algorithm 2: Feature pop operator

Consider a scenario where there are three samples in a local region called i, and these samples have the same class label as the central sample. Let's assume that the local feature subset obtained for this region is denoted as $f^{(i)}$, which consists of three features: f_1, f_3, and f_4. Additionally, the three samples from the overlapping region i have their respective local feature subsets: $f^{(1)} = \{f_1, f_3\}$, $f^{(2)} = \{f_2, f_3\}$ and $f^{(3)} = \{f_2, f_3, f_5\}$. To determine the importance of each feature in $f^{(i)}$, we sort them in ascending order based on their frequency of appearance in the above-mentioned feature subsets. This results in the sorted feature sequence $\{f_4, f_1, f_3\}$. Next, the feature pop operation is performed by removing features one by one from $f^{(i)}$ in the sorted sequence. The first feature to be removed is f_4, resulting in a new subset $\{f_1, f_3\}$. If the evaluation of this new subset, denoted as $rp^{(i)}(\{f_1, f_3\})$, yields a value that is lower than $rp^{(i)}(\{f_1, f_3, f_4\})$, then the feature push operator is stopped, and the updated local feature subset becomes $f^{(i)} = \{f_1, f_3\}$. If the evaluation does not satisfy the stopping criterion, we continue removing features one by one, considering them in the sorted sequence. For instance, f_1 is removed next, creating the subset $\{f_4, f_3\}$, which is then evaluated. This process continues until the stopping criterion is met. The algorithmic steps for the pop operator can be found in Algorithm 2.

4.3.3 Ensembles of Local Models

4.3.3.1 Genetic Operators

(1) Quick bit mutation:

In RP-LFS, we employ a rapid bit mutation technique that takes into consideration both the proportion of variant genes and the proportion of encoded genes. This
addresses the problem of an excessive number of 0-encoded genes being more prone
to switching to 1 during the evolutionary search process when the proportion of
selected features is relatively low, resulting in a slower convergence of evolutionary
search. The procedure commences with the assignment of a mutation rate μ, following which the number of selected genes is determined by Eq. (4.11) that randomly
selects G genes encoded as either 1 or 0. Finally, the selected genes' values are
flipped. Figure 4.4 illustrates this process using an individual chromosome with ten
genes, where a chromosome with four ones and six zeros has a mutation rate of 0.5.

$$G = \mu \times \min(count(0), count(1)) \tag{4.11}$$

(2) Network-inspired crossover:

The basic crossover operator does not favor either parent, where approximately half
of the genes originate from the first parent and the remaining genes come from the
second parent. However, selecting both parents randomly can affect the diversity
of the population and lead to the production of unfit offspring. Specifically, if the
genes of the two parents are highly similar, the offspring's genes will be identical to those of the parents. To overcome this issue, we draw inspiration from the
complex network theory [34] and establish a network of relationships to model the

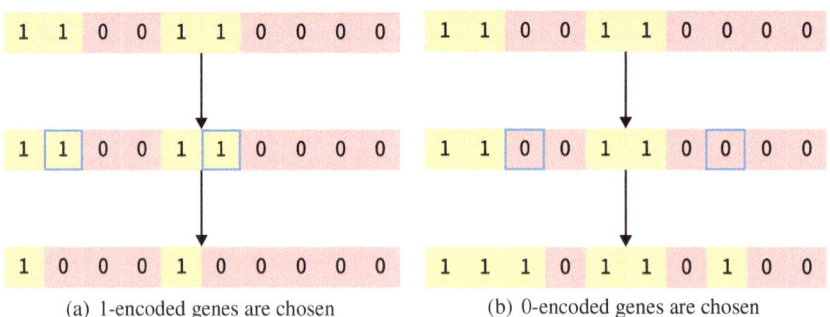

(a) 1-encoded genes are chosen   (b) 0-encoded genes are chosen

Fig. 4.4 Illustration of the proposed quick bit mutation strategy, where the number of flipped genes
is $G = 0.5 \times \min(4, 6) = 2$

similarity between one parent and the other individuals. Individuals with high similarity, i.e., with too many overlapping features, are removed from consideration, and the other parent is selected from individuals with low similarity. Our proposed network-inspired crossover operator follows the following process.

Step 1: Select a random individual as parent 1.

Step 2: Calculate the overlap of genetic properties between parent 1 and the remaining individuals $(C_1, \ldots, C_n)$.

Step 3: Standardize the overlap using Eq. (4.12) by taking into account the maximum (C_{max}) and minimum (C_{min}) values of similarity between parent 1 and the other individuals.

$$C\_normal(i) = \frac{C_i - C_{min}}{C_{max} - C_{min}} \quad i = 1, 2, \ldots, N \qquad (4.12)$$

Step 4: Exclude individuals that meet the condition C_normal(i)>rand(), resulting in M individuals with reduced similarity.

Step 5: Parent 2 is chosen based on the probability computed using Eq. (4.13).

$$p(j) = \frac{\exp^{C\_normal(j)}}{\sum_{m=1}^{M} \exp^{C\_normal(m)}} \quad j = 1, 2, \ldots, M \qquad (4.13)$$

Step 6: Randomly select half of the genetic material from the first parent and the remaining half from the second parent.

4.3.3.2 Classification

As multiple local feature subsets are generated by LFS, conventional classification methods that rely on a shared feature subset are not suitable. Moreover, to ensure a fair comparison and given our primary focus on the efficacy of LFS, we adopt the classification method employed in CLONALG-LFS [18] and LFSDC [20]. Specifically, in order to determine the label of a new sample $\mathbf{x}^p$, we measure the similarity $S(\mathbf{x}^p, c_k)$ between $\mathbf{x}^p$ and c_k based on the number of hyperspheres $Q^{(i)}$ with class label c_k that contain $\mathbf{x}^p$. The specific expressions for these measurements are as follows:

$$S\left(x^p, c_k\right) = \frac{\sum_{i \in Y_{c_k}} \psi_i \left(\mathbf{x}^p, r^{(i)}\right)}{\sum_{Y_{c_k}}} \qquad (4.14)$$

where the Y_{c_k} represents a collection of training samples that are classified as c_k, and $\psi(\cdot)$ is utilized to verify if the new sample $\mathbf{x}^p$ falls within the local region of $Q^{(i)}$ with the class label c_k, with the nearest neighboring sample belonging to the class label c_k. In case it satisfies the condition, $\psi(\cdot) = 1$; otherwise, $\psi(\cdot) = 0$. Upon calculating the similarity between x^p and all classes, the class label assigned to $\mathbf{x}^p$ is the one with the highest similarity.

4.3.3.3 The Overall Algorithm

We provide the pseudocode of RP-LFS in Algorithm 3. Firstly, the training set is partitioned into N local regions, following which multi-objective optimization is conducted on each region to determine the corresponding local feature subset. Upon completion of an iteration for all local regions, RFSS procedure is executed to update all feature subsets. RP-LFS operates iteratively until reaching the maximum iteration limit. Ultimately, with the optimal local feature subsets obtained for all N local regions, Eq. (4.14) is employed to classify the testing samples.

Input: Size of the training set N, maximum iteration T_{max} and testing samples;
Output: The last generation $\left\{ f(t+1)^{(i)} \right\}_{i=1}^{N}$ of all the local regions and labels of testing
    samples;
Initialize $t = 0$, Initialize population of all the local regions $\left\{ f(t)^{(i)} \right\}_{i=1}^{N}$;
while $t < T_{max}$ **do**
 | **while** $i < N$ **do**
 | | **for** each individual in population **do**
 | | | Calculate the fitness of an individual:
 | | | Apply Eq. (4.5) to compute $rp^{(i)}$;
 | | | Apply Eq. (4.6) to compute $pfs^{(i)}$;
 | | | Apply Eq. (4.9) to compute $rdm^{(i)}$.
 | | **end**
 | | Use Eq. (4.11) and Eq. (4.13) to produce new offsprings.
 | | Find the non-dominated set from all the individuals.
 | | Update the best member $f(t)^{(i)}$ from the non-dominated sets and perform the
 | | solution selection.
 | **end**
 | Update all the selected feature subsets $\left\{ f(t)^{(i)} \right\}_{i=1}^{N}$ using feature push operator in
 | Algorithm 1.
 | Update all the selected feature subsets $\left\{ f(t)^{(i)} \right\}_{i=1}^{N}$ using feature pop operator in
 | Algorithm 2.
end
Use Eq. (4.14) to perform classification and output the labels of testing samples.

Algorithm 3: RP-LFS

Our approach involves selecting only a single offspring solution from the two solutions generated by the conventional crossover operator. As illustrated in Fig. 4.5, we randomly select the gene values from both parent chromosomes of length 10 to construct the offspring's chromosome. The use of the network-inspired crossover operator enhances the efficiency of the offspring generated. This is due to the increased diversity in the genes of the new offspring, stemming from the maximal dissimilarity between the parents' genes. This not only expands the search space but also mitigates the risk of falling into a local optimum.

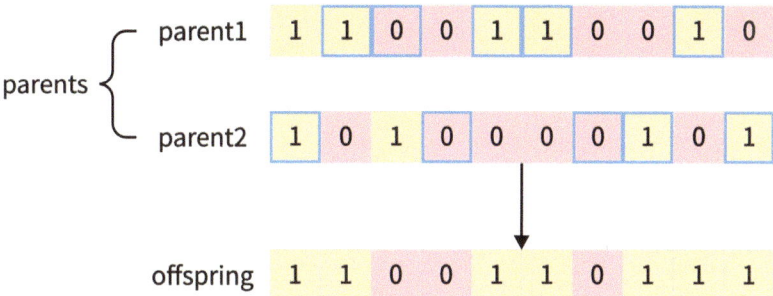

Fig. 4.5 Illustration of network-inspired crossover operator. The genes in the black border in the parents are those passed to the offspring

(1) Solution selection:

In order to effectively address the challenge of solution selection in multi-objective evolutionary algorithms (MOEAs), especially in practical applications, our proposed methodology aims to carefully choose a suitable solution from the set of Pareto optimal solutions generated by the MOEA. We prioritize features that exhibit a stronger association with the classifier, while also considering the potential negative impact of sparsity on classification accuracy caused by an overly sparse selection of features. To accomplish this, our methodology places initial emphasis on the RP, as it directly influences the accuracy of the classification. In cases where multiple solutions possess the same RP, we employ the region-based distance to evaluate the overall generality of the model. Additionally, the proportion of selected features is taken into account. To elaborate further, we first identify a subset of solutions with the lowest RP. From within this subset, we select the solution with the smallest region-based distance as a means to prioritize the model's generality. Lastly, the solution with the fewest selected features is chosen as the final solution. The decision to prioritize region-based distance over the proportion of selected features will be comprehensively analyzed in Sect. 4.4.7.

4.4 Results and Analysis

4.4.1 Experimental Design

4.4.1.1 Datasets

To evaluate the performance of the proposed RP-LFS method, we conducted experiments using a variety of datasets. Firstly, we selected 11 datasets from the UCI Repository, which are commonly used for classification tasks. Table 4.1 displays the

Table 4.1 The detail of the UCI datasets

Classes dataset	Instances	Classes	Features
Breast cancer	699	2	9
ORL	400	40	1024
Yale	165	15	1024
Wine	178	3	13
Musk1	476	2	166
Sonar	208	2	60
Iris	150	3	4
Vehicle	846	4	18
Ionosphere	351	2	34
Heart statlog	270	2	13
Diabetes	768	2	8

Table 4.2 The detail of the high-dimensional datasets

Dataset	Instances	Classes	Features	%Largest class	%Smallest class
DLBCL	77	2	5,469	75	25
9Tumor	60	9	5,726	15	3
Brain Tumor 1	90	5	5,920	67	4
Prostate	102	2	10,509	51	49
SRBCT	83	4	2,308	35	13
Leukemia 1	72	3	5,327	53	13
Lung Cancer	203	5	12,600	68	3
11Tumor	174	11	12,533	16	4
Leukemia 2	72	3	11,225	39	28

details of these datasets, including the number of instances, features, and classes. The number of features ranges from 4 to 1024, and the dataset sizes are significantly larger than the number of features.Furthermore, we applied the proposed algorithm to high-dimensional gene datasets, specifically employing nine benchmark gene datasets [23]. Table 4.2 provides information about these datasets, including the number of features, samples, classes, as well as the proportion of the smallest and largest classes. It is worth noting that most of the datasets have a limited number of samples, and there is a notable imbalance between the largest and smallest classes. All the datasets used in our experiments are available at https://github.com/primekangkang/Genedata.

4.4.1.2   Experimental Settings

In order to ensure the reliability and address any potential bias in the results of the FS (Feature Selection) method, we employ a 10-fold cross-validation technique to create the training and testing sets [4]. It is important to note that, as our proposed algorithm does not generate training accuracy, no validation set is utilized during the LFS (Local Feature Selection) process. The utilization of the training data varies slightly between the UCI dataset and the high-dimensional dataset, which will be discussed in more detail below. To ensure a fair comparison, we set the population size to 20, which is the same as that of the CLONALG-LFS [20] algorithm previously used in UCI datasets. Leveraging the effectiveness of our genetic operators and the regional feature sharing strategy, we choose to perform only half of the maximum number of iterations. For the first eight datasets, 50 iterations are conducted, while for the last three datasets, 100 iterations are performed. The relationship between the number of iterations and the RP is illustrated in Figs. 4.6 and 4.7. It can be observed that relatively stable results are achieved when the number of iterations approaches 100.Considering the intricacy of high-dimensional features, we set the number of populations to 100. After 120 iterations, a relatively stable solution can be obtained. Therefore, we set 120 iterations for all high-dimensional data (Table 4.3).

As it is widely acknowledged, during the iterative process in feature selection, the number of effective features tends to decrease significantly, resulting in a small amount of remaining features (typically between 15 to 40 for high-dimensional datasets). To ensure the gradual and accurate removal of irrelevant features via the flip operation, a small mutation rate, denoted as μ, is recommended. Our experiments have shown that a value of $\mu = 0.1$ is optimal.When dealing with UCI and high-dimensional datasets with a limited number of samples, the number of samples sharing the same label within any given local region is also small. In particular, when K (the number of nearest neighbors considered for classification) is large, the limited number of samples with the same label may result in misclassification, especially

Table 4.3 Parameter settings

Parameters	High-dimensional datasets	UCI datasets
Iteration T_{max}	120	50
Population size	100	20
The number of crossover	The same as the population size	
The number of mutation	The same as the population size	
The impurity level γ	0.2	
Mutation rate μ	0.1	
The number of the nearest samples K	1	
Divisions along each objective axis	27	

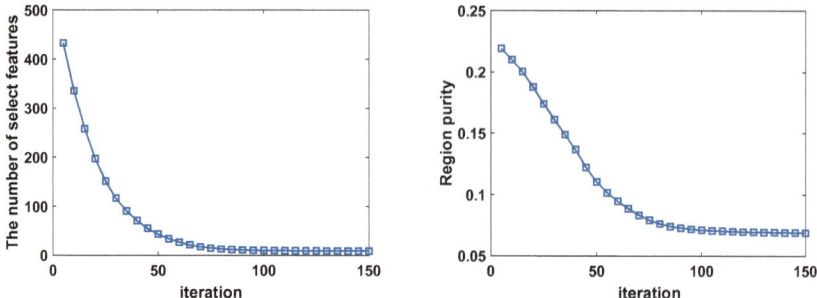

Fig. 4.6 Convergence of the RP and the number of features during the iteration process for UCI dataset (Yale)

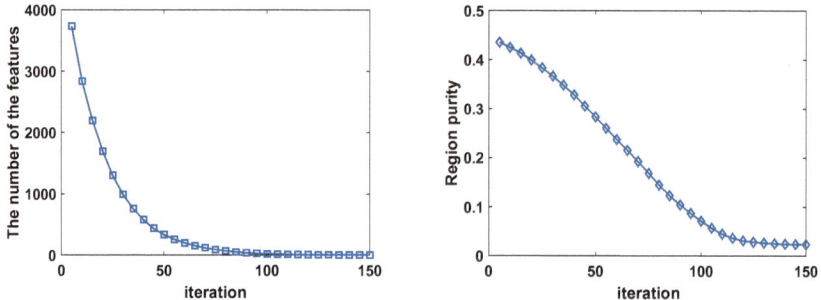

Fig. 4.7 Convergence of the RP and the number of features during the iteration process for a high-dimensional dataset (Prostate)

for samples lying on the boundary. In contrast, most state-of-the-art feature selection (FS) methods adopt a value of $K = 1$ in their experiments. To ensure the effectiveness of our method when dealing with class imbalance and datasets with few samples, and to facilitate fair comparisons, we also set $K = 1$ in our experiments. Regarding the number of divisions along each objective axis, we follow the settings suggested in [29]. As for the impurity level for local region division, we adopt the settings recommended in [18].

4.4.2 Comparison with State-of-the-Art Methods

4.4.2.1 Comparison of UCI Datasets

To assess the efficacy of our proposed RP-LFS algorithm, we conducted a comprehensive evaluation by comparing its classification accuracy against seven widely recognized filter-based feature selection (FS) algorithms and two LFS algorithms on UCI datasets. The specific algorithms used for comparison are as follows: mRMR [35],

$VMI_{pairwise}$ [36], JMIM [35], Simba [37], ReliefF [38], Information Gain [39], Information Gain Ratio [39], CLONAL-LFS [20], and LFSDC [19]. As the algorithms under consideration are filter-based FS and LFS-based, we utilized a 10-fold cross-validation approach, where the training and testing sets were partitioned accordingly. Importantly, no validation set was employed in this evaluation. The code for the first eight algorithms was readily available, and we executed the code to obtain the respective results. As for the LFSDC algorithm, we obtained the code from its authors and extended it to accommodate a multi-class scenario. It is noteworthy that all parameter settings adhered to the recommendations provided in the corresponding papers.

We performed a statistical Wilcoxon significance test [40] at a 5% significance level. Our proposed RP-LFS serves as the reference algorithm, with the following interpretations: "+" indicates that RP-LFS performs significantly better than the compared algorithm, "−" indicates that RP-LFS performs worse, and "=" indicates similar results between the two algorithms. Since the evolutionary algorithm is a stochastic method, 30 independent runs were conducted for each method using different random seeds. This resulted in a total of 300 runs (30 runs x 10 fold cross-validation) for each method on each dataset. All numerical results were obtained using a PC equipped with an Intel Core i7-7700 CPU @ 3.6GHz and 32GB of memory.

The statistical results can be found in Tables 4.4 and 4.5. The tables present the outcomes of the experimental evaluation, showcasing the smallest feature subset and the best classification accuracy for each dataset. The optimal results are indicated in bold, while the second best results are underlined. After calculating the average values across all datasets, it is evident that RP-LFS demonstrates remarkable performance. With an average classification accuracy of 89.59% and a minimal feature size of 9.66, RP-LFS outperforms other algorithms. Additionally, RP-LFS achieves superior results on seven out of the eleven datasets. This algorithm's effectiveness can be attributed to its ability to consider the performance of each local region, making it particularly efficient for multi-classification and high-dimensional datasets. Furthermore, RP-LFS is not limited by the distance function. By selecting the optimal feature subset for each local region, both the final performance and the number of selected features are significantly improved.

4.4.2.2 Comparison of High-Dimensional Datasets

To evaluate the performance of feature selection algorithms for high-dimensional datasets, we assess the classification outcomes and significance test results of five state-of-the-art global feature selection methods. These include three PSO-based algorithms, namely PPSO [43], VLPSO [23], and VLPSO-LFS [23], as well as two NSGA-based algorithms, NSGAIII-FS [29] and PS-NSGA [29]. Additionally, we consider two representative LFS methods, namely CLONAL-LFS [20] and LFSDC [19].

Considering that these five EA-based methods are wrapper-based feature selection (FS) techniques, it is important to calculate the training accuracy using the validation set. However, LFS methods do not require the use of a validation set. In order to ensure

Table 4.4 Average classification accuracies and significance test of the different algorithms in UCI datasets

Dataset	mRMR	JMIM	VMI	Simba	ReliefF	IG	IGR	CLONALG-LFS	LFS-DC	RP-LFS
Vehicle	65.75(+)	64.31(+)	64.96(+)	66.93(+)	64.59(+)	64.78(+)	64.59(+)	63.55(+)	67.74(+)	**74.62**
Iris	**96.65**(=)	95.93(=)	95.96(=)	95.65(=)	95.96(=)	95.96(=)	95.96(=)	92.00(+)	94.00(+)	96.27
Diabetes	**75.66**(=)	75.06(=)	75.46(=)	75.53(=)	75.06(=)	75.06(=)	75.14(=)	73.38(+)	75.39(=)	75.51
Breast Cancer	96.50(=)	**96.70**(=)	96.50(=)	96.50(=)	96.57(=)	96.54(=)	96.68(=)	95.48(=)	96.28(=)	96.42
Musk1	85.33(+)	84.76(+)	86.49(+)	84.37(+)	84.83(+)	84.68(+)	86.00(+)	86.89(+)	79.33(+)	**91.11**
Wine	97.77(=)	97.66(=)	97.67(=)	97.66(=)	97.30(=)	97.60(=)	97.74(=)	97.78(=)	94.84(-)	**97.98**
ORL	91.31(+)	91.17(+)	92.63(+)	93.72(+)	90.91(+)	90.80(+)	90.75(+)	87.50(+)	96.65(=)	**96.75**
Ionosphere	91.26(+)	90.00(+)	91.69(+)	91.30(+)	91.61(+)	91.72(+)	90.71(+)	**94.34**(=)	91.79(+)	94.12
Heart statlog	83.11(=)	83.30(=)	83.16(=)	82.75(=)	82.75(=)	83.31(=)	83.31(=)	**83.33**(=)	80.24(+)	83.11
Sonar	79.14(+)	82.10(+)	83.48(+)	82.85(+)	80.64(+)	79.26(+)	79.26(+)	87.97(+)	79.66(+)	**89.15**
Yale	67.93(+)	67.17(+)	70.73(+)	70.97(+)	67.53(+)	66.64(+)	66.28(+)	73.33(+)	81.66(+)	**90.08**
AVG	84.8	84.37	85.34	85.29	84.34	84.21	84.22	85.06	85.23	**89.56**
WIN	2	1	0	0	0	0	0	2	0	6

Table 4.5 The average number of the selected features in UCI datasets

Dataset	mRMR	JMIM	VMI	Simba	ReliefF	IG	IGR	CLONALG-LFS	LFSDC	RP-LFS
Vehicle	15	17	7	10	18	16	18	5	4.43	**2.95**
Iris	2	3	3	2	3	3	3	**1**	1.51	1.92
Diabetes	4	8	5	5	8	8	6	**1**	3.39	2.94
Breast cancer	9	7	9	9	7	6	8	**2**	4.29	4.95
Musk1	82	163	87	155	162	115	88	22	**15.87**	24.43
Wine	7	5	8	6	13	6	10	**2**	3.91	5.85
ORL	436	311	383	184	769	774	769	35	93.13	**29.73**
Ionosphere	15	5	4	20	9	8	14	11	**3.58**	5.29
Heart statlog	**3**	11	11	13	13	9	9	5	4.43	4.95
Sonar	58	25	20	24	40	59	59	11	**5.38**	12.90
Yale	227	578	157	113	273	918	988	15	67.29	**10.3**
AVG	78	103	63.09	49.18	119.55	174.73	179.27	10	18.846	**9.66**
WIN	1	0	0	0	0	0	0	4	3	3

a fair comparison, we initially split the dataset into training, validation, and testing sets, followed by performing a 10-fold cross validation as recommended in [23, 29]. To ensure equal treatment between the EA-based methods and LFS methods in our experiment, we disregard the validation set so that an equal number of samples are used for training. We use the provided source code from the respective authors to obtain the results for each approach. Consequently, we conduct a total of 300 runs (30 runs x 10-fold cross validation) for each method on each dataset, and record the balanced classification accuracy on the test data.

The findings from our experimentation can be found in Tables 4.6 and 4.7 , illustrating the efficacy of our proposed RP-LFS technique when applied to high-dimensional data. We further conducted a statistical Wilcoxon significance test at a 5% significance level. The average number of selected features in LFS reflects the quantity of features chosen across all local regions. Our approach surpasses state-of-the-art algorithms, exhibiting superior performance in both classification accuracy and feature subset size. Notably, RP-LFS outperforms other algorithms on seven out of nine datasets. Across all datasets, RP-LFS achieves an average classification accuracy of 89.04% and a feature subset size of 22.83, outperforming all other tested algorithms.

CLONALG-LFS and LFSDC lack the ability to guarantee that the selected features are region-based because their local feature selection processes are independent of the region segmentation in classification. In contrast, RP-LFS integrates regional segmentation and LFS into the same stage, resulting in a selected feature subset that aligns more accurately with regional characteristics and takes into account the feature interaction across multiple local regions. This approach ultimately yields superior classification accuracy.

Table 4.6 Average classification accuracies and significance test of the different algorithms in high-dimensional datasets

Dataset	NSGAIII-FS	PPSO	VLPSO	VLPSO-LS	PS-NSGA	CLONALG-LFS	LFSDC	RP-LFS
11Tumor	74.09(+)	76.83(+)	80.81(+)	82.81(+)	83.94(+)	75.79(+)	83.15(+)	**86.29**
Leukemia1	84.44(+)	94.37(+)	93.31(+)	93.75(+)	91.98(+)	82.92(+)	93.99(+)	**95.23**
SRBCT	91.13(+)	95.78(+)	99.67(=)	99.75(=)	96.35(+)	83.47(+)	99.58(=)	**99.92**
DLBCL	83.33(+)	86.22(+)	86.51(+)	96.13(+)	87.02(+)	80.83(+)	91.13(+)	**96.67**
Prostate	84.72(+)	91.82(+)	89.82(+)	92.58(+)	89.44(+)	82.36(+)	88.55(+)	**94.40**
Lung Cancer	74.12(+)	79.38(+)	89.47(+)	90.17(+)	88.48(+)	88.59(+)	90.36(+)	**92.30**
Leukemia2	88.20(+)	96.74(+)	91.56(+)	95.39(+)	92.22(+)	87.86(+)	94.78(+)	**98.37**
Brain_Tumor1	76.16(+)	74.40(+)	71.19(+)	75.54(+)	73.81(+)	72.89(+)	**77.43** (=)	77.28
9Tumor	48.43(+)	59.28(+)	55.11(=)	56.78(+)	58.30(+)	37.28(+)	57.85(+)	**60.38**
AVG	78.29	83.87	84.16	86.99	84.62	78.24	86.31	**89.04**
WIN	0	0	0	0	0	0	1	8

Table 4.7 The average number of the selected features in high-dimensional datasets

Dataset	NSGAIII-FS	PPSO	VLPSO	VLPSO-LS	PS-NSGA	CLONALG-LFS	LFSDC	RP-LFS
11Tumor	5578.3	167.0	246.7	367.4	338.3	6317.4	178.2	**39.34**
Leukemia1	2085.2	80.4	54.7	59.3	**16.2**	2697.2	141.2	20.84
SRBCT	789.7	108.5	49.1	71.4	18.6	1181.8	113.8	**15.52**
DLBCL	2186.5	44.0	48.1	59.9	**9.4**	2774.6	166.2	15.19
Prostate	4556.1	65.6	26.4	56.4	65.0	5300.1	165.21	**10.73**
Lung Cancer	5489.4	203.0	176.1	242.9	107.6	6350.0	188.3	**44.81**
Leukemia2	4990.9	86.7	35.2	61.2	28.4	5661.0	154.8	**17.41**
Brain_Tumor1	2335.2	73.4	26.8	102.1	57.8	2995.9	172.4	**21.55**
9Tumor	2264.6	118.1	44.2	61.9	194.8	2900.5	186.8	**19.11**
AVG	3363.99	105.19	78.59	120.28	92.9	4019.8	162.99	**22.83**
WIN	0	0	0	0	2	0	0	7

4.4.3   The Effectiveness of RP

In order to validate the effectiveness of our proposed RP algorithm as an objective, we conducted a systematic comparison between RP and the traditional distance function in the context of local feature selection (LFS). To ensure a fair comparison, we eliminated the influence of the region-based distance metric present in RP-LFS. Instead, we compared RP with two objectives (RP and proportion of selected features) against the LFS algorithm using two distance-based functions (maximizing inter-class distance and minimizing intra-class distance). Both methods were implemented using the same NSGA II-based optimization algorithm, which includes the network-based crossover operator, quick bit mutation operator, and RFSS. For clarity, we refer to these two methods as RP-NSGAII and DISTANCE-NSGAII, respectively. To evaluate their performance, we executed each algorithm 30 times with 10-fold cross validation for each dataset. For visualization purposes, we selected the results from two representative datasets. As illustrated in Figs. 4.8 and 4.9, the iteration process demonstrated that RP-NSGAII achieved significantly higher classification accuracy with fewer selected features compared to distance-NSGAII. While the fast convergence of selected features can be attributed to RP-NSGAII's objective function considering the proportion of selected features, the improvement in classification accuracy confirms that our proposed RP algorithm is more effective than commonly-used distance-based functions in enhancing LFS accuracy. Furthermore, comparing the results of RP-LFS and RP-NSGAII reveals that optimizing all three objectives leads to improved classification performance, while introducing the region-based distance metric enhances the model's generality. This is the rationale behind applying three objectives in our RP-LFS approach.

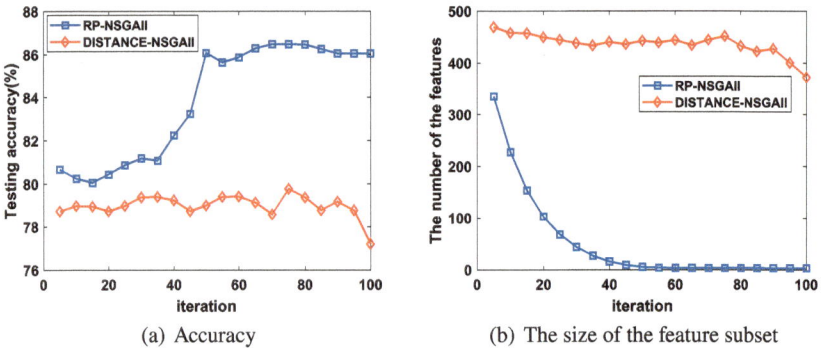

(a) Accuracy   (b) The size of the feature subset

Fig. 4.8 Comparison between the distance-based NSGAII and RP-based NSGAII on Yale, a typical multi-class dataset

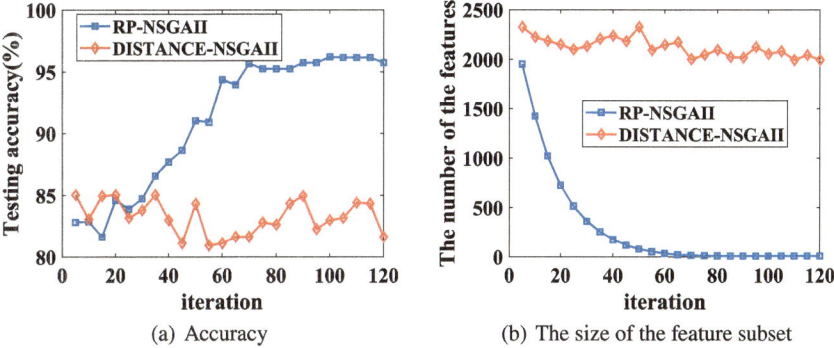

Fig. 4.9 Comparison between the distance-based NSGAII and RP-based NSGAII, a high-dimensional dataset

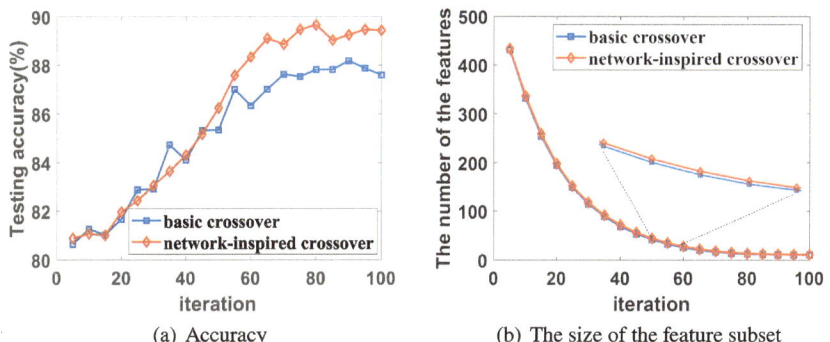

Fig. 4.10 Comparison between network-inspired crossover and the basic crossover on Yale (a typical multi-class dataset)

4.4.4 Impact of the Network-Inspired Crossover

To authenticate the efficacy of the network-inspired crossover, we assess RF-LFS in comparison with a network-inspired crossover operator which encompasses only the basic crossover operator. Without any loss of generality, we present the comparative outcomes on a typical multi-class dataset named Yale, due to the fact that the general behavior for other datasets is nearly equivalent. As depicted in Fig. 4.10, the pace of feature reduction accomplished by the basic crossover surpasses our network-inspired crossover in the initial stages of the iterative process, as the network-inspired crossover is tailored to find diverse parents so as to produce diverse offspring solutions. However, in the later stages, the network-inspired crossover manages to produce comparatively higher accuracy as the network's ability to enhance population diversity helps to mitigate the chances of the algorithm getting trapped into the local optima.

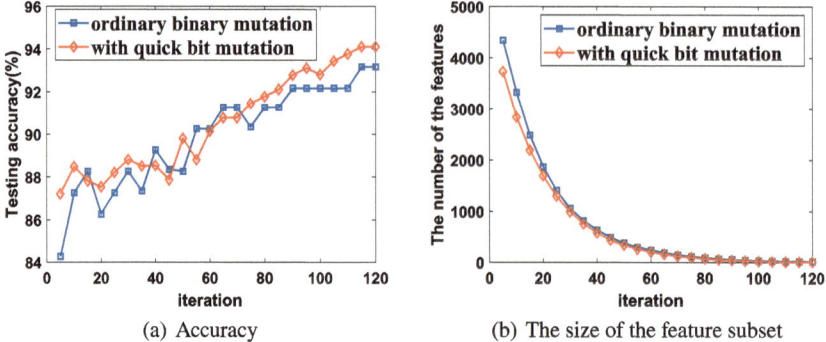

(a) Accuracy   (b) The size of the feature subset

Fig. 4.11 Comparison between the quick bit mutation and the ordinary binary mutation on Prostate, a typical high-dimensional dataset

4.4.5   Impact of Quick Bit Mutation

In order to evaluate the efficacy of the rapid bit mutation operator, a comparison is conducted between our algorithm incorporating rapid bit mutation and another algorithm utilizing ordinary binary mutation. The graphical illustration presented in Fig. 4.11 demonstrates that, as the iterative process progresses, the implementation of rapid bit mutation facilitates the algorithm in escaping local optima and attaining higher classification accuracy. Additionally, rapid bit mutation exhibits a faster convergence in terms of the number of selected features, in contrast to the utilization of ordinary binary mutation.

4.4.6   Impact of RFSS

The primary objective of the regional feature sharing strategy is to enhance feature selection performance by leveraging the shared features among local regions.

Observing Fig. 4.12, it is evident that, during the initial iterations, both scenarios, i.e., with and without the regional feature sharing strategy, exhibit a similar rate of feature reduction. However, as the process continues, the scenario without the regional feature sharing strategy tends to result in a reduced number of features. Conversely, employing the regional feature sharing strategy effectively preserves some valuable features, thereby ensuring a higher classification accuracy from start to finish. Additionally, this strategy capitalizes on the feature sharing within the overlapping regions, resulting in an expedited attainment of the convergence state for the optimization objective.

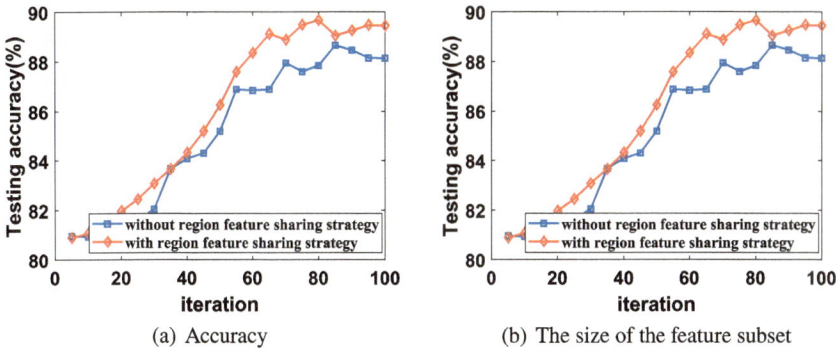

Fig. 4.12 Comparison between the algorithm with RFSS and without RFSS on Yale, a typical multi-class dataset

4.4.7 Studies on Solution Selection

As previously noted, the selection of an optimal solution from the non-dominated set is a critical task for multi-Objective evolutionary algorithms (MOEAs). In our research, we aim to explore the relationship between local feature selection and classifier, and thus employ RP as our primary preference. However, in many instances, multiple solutions possess equivalent RP values, necessitating the use of additional indicators to identify the preferred solution. For this purpose, we compare the region-based distance and the proportion of selected features to determine the secondary metric. After applying RP, we conduct a comparison of these two metrics, respectively. Unlike traditional feature selection, where only a set of solutions is generated for multi-objective optimization, local feature selection (LFS) generates multiple subsets of feature sets, making direct analysis challenging. Consequently, we validate the efficiency of the solution selection strategy using testing accuracy since RP-LFS lacks training accuracy. It is essential to note that we do not base the preferred solution on testing accuracy, which may cause biases, but on the three objectives utilized in training. As illustrated in Fig. 4.13, the use of the region-based distance metric can significantly enhance the classification accuracy to a certain extent. Theoretically, the region-based distance metric is directly related to classification accuracy, and too few features lead to lower accuracy. Thus, selecting the region-based metric before the proportion of selected features is a more reasonable choice.

4.4.8 Comparison Between Original NSGA-III and Our Improved NSGA-III

In order to assess and compare the performance of the improved NSGA-III and the original NSGA-III in the context of local feature selection (LFS), we make use of

Fig. 4.13 Comparison
between the region-based
distance function and the
proportion of the selected
features as the the second
indicator in solution
selection strategy

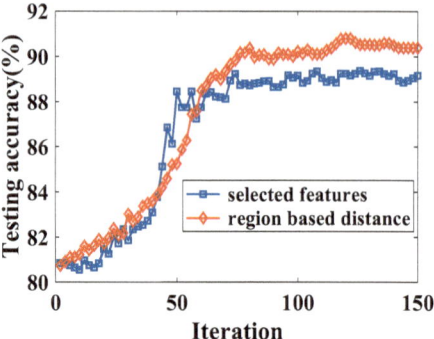

the original NSGA-III in place of the improved NSGA-III within the framework of
RP-LFS, which we refer to as RP-NSGAIII. The obtained Pareto set is then evaluated
using two performance metrics as outlined below:

4.4.8.1 Hypervolume (HV)

In relation to the HV [41] metric, all the objective values obtained by each algorithm
are normalized based on the ideal and nadir points, resulting in a normalized scale
ranging from 0 to 1 in each objective direction. The ideal and nadir points are specifically defined as $(0, 0, 0)$ and $(1.1, 1.1, 1.1)$, respectively. Subsequently, the reference
point for HV is set to $(1, 1, 1)$. Consequently, if a Pareto optimal set A is determined
to be superior to set B, the HV value of A will be higher in comparison to that of B.

4.4.8.2 Inverted Generational Distance (IGD)

The use of the IGD [42] metric enables the simultaneous evaluation of the convergence and distribution performance of a Pareto optimal set. Despite its significance,
the evaluation of feature selection, being a discrete and complex problem, does not
allow for the exact computation of the Pareto frontier. As such, the final populations
from both algorithms were collected and combined into a joint population to ensure a
fair comparison. Subsequently, the joint population underwent non-dominance ranking, and the first non-dominance frontier was retained as a reference point to calculate
IGD. Lower IGD values indicate superior algorithm performance, in contrast to the
HV metric.

In the specific case of RP-LFS and RP-NSGAIII, as shown in Table 4.8, RP-LFS
performs slightly better than RP-NSGAIII in terms of both HV and IGD. This suggests that the improved NSGA-III used in RP-LFS yields Pareto solutions that exhibit
better convergence and distribution when applied to the LFS problem, compared to
the original NSGA-III.

Table 4.8 Comparison between the original NSGA-III and the improved NSGA-III for RP-LFS

DataSet	Method	IGD (Average/Std)	HV (Average/Std)
11Tumor	RP-NSGAIII	1.4400/0.0060	0.7650/0.0060
	RP-LFS	**1.4461/0.0043**	**0.7748/0.0072**
Leukemia1	RP-NSGAIII	1.4379/0.0078	0.8106/0.0102
	RP-LFS	**1.4410/0.0058**	**0.8119/0.0048**
SRBCT	RP-NSGAIII	1.4432/0.0085	0.8431/0.0054
	RP-LFS	**1.4441/0.0056**	**0.8522/0.0036**
DLBCL	RP-NSGAIII	0.7564/0.0164	1.4153/0.0036
	RP-LFS	**1.4184/0.0042**	**0.7582/0.0157**
Prostate	RP-NSGAIII	1.3652/0.0059	0.5858/0.0157
	RP-LFS	**1.3679/0.0044**	**0.5938/0.0134**
Lung cancer	RP-NSGAIII	1.4167/0.0048	0.6395/0.0092
	RP-LFS	**1.4195/0.0008**	**0.6561/0.0072**
Leukemia2	RP-NSGAIII	1.4344/0.0041	0.8175/0.0099
	RP-LFS	**1.4366/0.0047**	**0.8214/0.0109**
Brain_Tumour1	RP-NSGAIII	1.4222/0.0055	0.6991/0.0152
	RP-LFS	**1.4265/0.0053**	**0.7002/0.0183**
9Tumors	RP-NSGAIII	1.4385/0.0083	0.7443/0.0184
	RP-LFS	**1.4442/0.0084**	**0.7692/0.0078**

4.4.9 Discussion

4.4.9.1 Three Objectives, Two Objectives and LFSDC

To showcase the efficacy of region-based distance, we examine RP-NSGAII, incorporating region-based distance as one objective, and the proportion of selected features as the other. Table 4.9 illustrates the number of features and the accuracy achieved by three algorithms, namely LFSDC, RP-NSGAII, and RP-LFS, on a high-dimensional dataset. It is evident that RP-NSGAII yields the lowest number of features across all datasets. Though RP-LFS surpasses RP-NSGAII with respect to accuracy, both algorithms demonstrate the potential of region-based distance in enhancing feature selection's generalization ability, thereby reinforcing our preference for distance-based selection criteria in Sect. 4.4.7. In terms of accuracy, RP-NSGAII outperforms LFSDC in DLBCL, Leukemia1, Prostate, Leukemia2, and 11Tumor, corroborating the effectiveness of our proposed RP fitness function and showcasing the capabilities of our two-objective algorithm.

Table 4.9 Comparison of the performance of our algorithm in two objectives and three objectives with LFSDC

DataSet	#Method	#Feature	Accuracy (std)
11Tumor	RP-NSGAII	**7.16**	85.59(3.17)
	LFSDC	178.2	83.15(2.32)
	RP-LFS	39.34	**86.29(3.29)**
Leukemia1	RP-NSGAII	**4.58**	95.11(1.44)
	LFSDC	141.2	93.99(2.91)
	RP-LFS	20.84	**95.23(1.27)**
SRBCT	RP-NSGAII	**3.26**	99.16(0.63)
	LFSDC	113.8	99.58(0.59)
	RP-LFS	15.52	**99.92(0.11)**
DLBCL	RP-NSGAII	**5.95**	95.17(2.53)
	LFSDC	166.2	91.13(2.88)
	RP-LFS	15.19	**96.67(1.08)**
Prostate	RP-NSGAII	**7.21**	93.00(3.17)
	LFSDC	165.21	88.55(2.32)
	RP-LFS	10.73	**94.40(2.32)**
Lung cancer	RP-NSGAII	**14.62**	89.60(2.18)
	LFSDC	188.3	90.36(1.53)
	RP-LFS	44.81	**92.30(1.15)**
Leukemia2	RP-NSGAII	**4.06**	98.33(1.85)
	LFSDC	154.8	94.78(1.85)
	RP-LFS	17.41	**98.37(1.27)**
BrainTumour1	RP-NSGAII	**8.59**	70.17(3.87)
	LFSDC	172.4	**77.43(3.52)**
	RP-LFS	22.55	77.28(3.33)
9Tumors	RP-NSGAII	**4.03**	50.44(5.37)
	LFSDC	186.8	57.85(4.71)
	RP-LFS	19.11	**60.38(2.39)**
Mean	RP-NSGAII	**6.61**	86.61
	LFSDC	162.99	86.31
	RP-LFS	22.83	**89.04**

4.4.9.2 Computational Time Analysis

In Fig. 4.14, taking into account the varying number of iterations utilized in the compared methods, we have calculated the average computational time for each iteration in order to ensure a fair comparison across all FS methods. Given that RP-LFS consists of N local regions (where N represents the number of training samples), it necessitates the use of N models. By employing algorithm parallelization, we record the computational time, and subsequently derive the average time for each

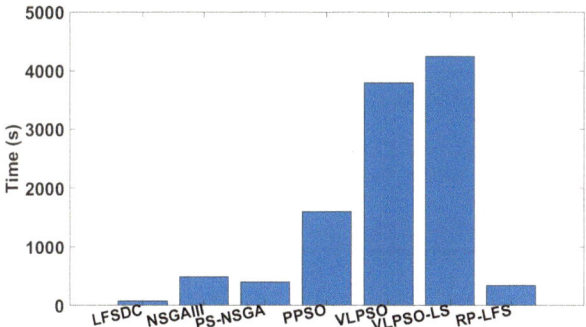

Fig. 4.14 Comparison of the running time

iteration. Remarkably, RP-LFS demonstrates significantly lower time consumption when compared to other global feature selection approaches, particularly, it only requires 10% of the time required by VLPSO-LS.

It is important to highlight that optimizing multiple local regions simultaneously is still a complex task, which presents higher complexity compared to LFS-DC. Additionally, achieving complete independence and parallelization of solutions can prove difficult due to the dynamics of these local regions, ultimately requiring an evolutionary optimization approach. Consequently, future research should focus on developing a suitable parallelization strategy to address this challenge.

References

1. Zhou, Y., Qiu, Y., Kwong, S.: Region Purity-based Local Feature Selection: A Multi-Objective Perspective. IEEE Trans. Evol. Comput, vol. 27, no. 4, pp. 787–801 (2023)
2. John, G.H., Kohavi, R., Pfleger, K.: Irrelevant features and the subset selection problem. In: Machine Learning Proceedings 1994, pp. 121–129 (1994)
3. Whitney, A.W.: A direct method of nonparametric measurement selection. IEEE Trans. Comput. **100**(9), 1100–1103 (1971)
4. Kohavi, R., et al.: A study of cross-validation and bootstrap for accuracy estimation and model selection. In: Ijcai, vol. 14, no. 2, pp. 1137–1145 (1995). (Montreal, Canada)
5. Reunanen, J.: Overfitting in making comparisons between variable selection methods. J. Mach. Learn. Res. **3**, 1371–1382 (2003)
6. Xue, B., Zhang, M., Browne, W.N.: Particle swarm optimization for feature selection in classification: a multi-objective approach. IEEE Trans. Cybern. **43**(6), 1656–1671 (2012)
7. Chen, Q., Xue, B., Zhang, M.: Genetic programming for instance transfer learning in symbolic regression. IEEE Trans. Cybern. (2022). https://doi.org/10.1109/TCYB.2020.2969689
8. Wang, P., Xue, B., Liang, J., Zhang, M.: Multiobjective differential evolution for feature selection in classification. IEEE Trans. Cybern. (2021)
9. De Stefano, C., Fontanella, F., Cristina Marrocco, A., Di Freca, S.: A GA-based feature selection approach with an application to handwritten character recognition. Pattern Recogn. Lett. **35**, 130–141 (2014)
10. Chen, K., Xue, B., Zhang, M., Zhou, F.: An evolutionary multitasking-based feature selection method for high-dimensional classification. IEEE Trans. Cybern. (2020)

11. Wang, Z., Li, M., Li, J.: A multi-objective evolutionary algorithm for feature selection based on mutual information with a new redundancy measure. Inf. Sci. **307**, 73–88 (2015)
12. Tabakhi, S., Moradi, P.: Relevance-redundancy feature selection based on ant colony optimization. Pattern Recogn. **48**, 2798–2811 (2015)
13. Tran, B., Xue, B., Zhang, M.: Genetic programming for multiple-feature construction on high-dimensional classification. Pattern Recogn. **93**, 404–417 (2019)
14. Zaretalab, A., Hajipour, V., Sharifi, M., Shahriari, M.R.: A knowledge-based archive multi-objective simulated annealing algorithm to optimize series-parallel system with choice of redundancy strategies. Comput. Ind. Eng. (2015). https://doi.org/10.1016/j.cie.2014.11.008
15. Hancer, E, et al.: Pareto front feature selection based on artificial bee colony optimization. Inf. Sci. **422**(2018), 462–479
16. Zhang, Y., Gong, D., Cheng, J.: Multi-objective particle swarm optimization approach for cost-based feature selection in classification. IEEE/ACM Trans. Comput. Biol. Bioinf. (TCBB) **14**(1), 64–75 (2017). (IEEE Computer Society Press)
17. Dudek, G.: An artificial immune system for classification with local feature selection. IEEE Trans. Evol. Comput. **16**, 847–860 (2012)
18. Armanfard, N., Reilly, J.P., Komeili, M.: Local feature selection for data classification. IEEE Trans. Pattern Anal. Mach. Intell. **38**, 1217–1227 (2015)
19. Armanfard, N., Reilly, J.P., Komeili, M.: Logistic localized modeling of the sample space for feature selection and classification. IEEE Trans. Neural Netw. Learn. Syst. **29**, 1396–1413 (2017)
20. Wang, Y., Li, T.: Local feature selection based on artificial immune system for classification. Appl. Soft Comput. **87**, 105989 (2020)
21. Zhang, Q., Li, H.: MOEA/D: A multiobjective evolutionary algorithm based on decomposition. IEEE Trans. Evol. Comput. **11**(6), 712–731 (2007)
22. Deb, K., Jain, H.: An evolutionary many-objective optimization algorithm using reference-point-based nondominated sorting approach, part I: solving problems with box constraints. IEEE Trans. Evol. Comput. **18**(4), 577–601 (2013)
23. Tran, B., Xue, B., Zhang, M.: Variable-length particle swarm optimization for feature selection on high-dimensional classification. IEEE Trans. Evol. Comput. **23**(3), 473–487 (2019)
24. Wang, Y., Liu, B., Ma, Z., Wong, K.-C., Li, X.: Nature-inspired multiobjective cancer subtype diagnosis. IEEE J. Transl. Eng. Health Med. (2019). https://doi.org/10.1109/JTEHM.2019. 2891746
25. Nguyen, B.H., Xue, B., Andreae, P., Ishibuchi, H., Zhang, M.: Multiple reference points-based decomposition for multiobjective feature selection in classification: static and dynamic mechanisms. IEEE Trans. Evol. Comput. (2020). https://doi.org/10.1109/TEVC.2019.2913831
26. Khan, B., Johnstone, M., Hanoun, S., Lim, C.P., Creighton, D., Nahavandi, S.: Improved NSGA-III using neighborhood information and scalarization. IEEE Int. Conf. Syst. Man Cybern. (SMC) (2016). https://doi.org/10.1109/SMC.2016.7844702
27. Zhihua Cui, Yu., Chang, J.Z., Cai, X., Zhang, W.: Improved NSGA-III with selection-and-elimination operator. Swarm Evol. Comput. (2019). https://doi.org/10.1016/j.swevo.2019.05. 011
28. Zhu, Y., Liang, J., Chen, J., Ming, Z.: An improved NSGA-III algorithm for feature selection used in intrusion detection. Knowl. Based Syst. **116**, 74–85 (2017). (Elsevier)
29. Zhou, Y., Zhang, W., Kang, J., Zhang, X., Wang, X.: A problem-specific non-dominated sorting genetic algorithm for supervised feature selection. Inf. Sci. **547**, 841–859 (2021)
30. Yuan, X., Tian, H., Yuan, Y., Huang, Y., Ikram, R.M.: An extended NSGA-III for solution multi-objective hydro-thermal-wind scheduling considering wind power cost. Energy Convers. Manag. (2015). https://doi.org/10.1016/j.enconman.2015.03.009
31. Al-Sahaf, H., Zhang, M., Johnston, M., Verma, B.: Image descriptor: a genetic programming approach to multiclass texture classification. In: 2015 IEEE Congress on Evolutionary Computation (CEC), pp. 2460–2467 (2015)
32. Song, X.-F., Zhang, Y., Gong, D.-W., Gao, X.-Z.: A fast hybrid feature selection based on correlation-guided clustering and particle swarm optimization for high-dimensional data. IEEE Trans. Cybern. (2021)

33. Chen, K., Xue, B., Zhang, M., Zhou, F.: Correlation-guided updating strategy for feature selection in classification with surrogate-assisted particle swarm optimisation. IEEE Trans. Evol. Comput. (2021)
34. Erds, P., Rényi, A.: On the evolution of random graphs (1961)
35. Bennasar, M., Hicks, Y., Setchi, R.: Feature selection using joint mutual information maximisation. Expert Syst. Appl. **42**, 8520–8532 (2015)
36. Gao, S., Ver Steeg, G., Galstyan, A.: Variational information maximization for feature selection. NIPS (2016)
37. Gilad-Bachrach, R., Navot, A., Tishby, N.: Margin based feature selection-theory and algorithms. Presented at the (2004)
38. Kononenko, I.: Estimating attributes: analysis and extensions of RELIEF. In: European Conference on Machine Learning, pp. 171–182 (1994)
39. Ross Quinlan, J.: Induction of decision trees. Mach. Learn. **1**, 81–106 (1986)
40. Derrac, J., García, S., Molina, D., Herrera, F.: A practical tutorial on the use of nonparametric statistical tests as a methodology for comparing evolutionary and swarm intelligence algorithms. Swarm Evol. Comput. **1**, 3–18 (2011)
41. While, L., Hingston, P., Barone, L., Huband, S.: A faster algorithm for calculating hypervolume. IEEE Trans. Evol. Comput. (2006). https://doi.org/10.1109/TEVC.2005.851275
42. Coello, C.A.C., Cortés, N.C.: Solving multiobjective optimization problems using an artificial immune system. Genetic Program. Evolvable Mach. **6**, 163–190 (2005)
43. Tran, B., Xue, B., Zhang, M.: A new representation in PSO for discretization-based feature selection. IEEE Trans. Cybern. **48**(6), 1733–1746 (2017). (IEEE)

Chapter 5
Deep Neural Network Based Hybrid Feature Selection

Abstract With the widespread adoption and proliferation of Internet of Things (IoT) technology, the human recognition activity (HAR) utilizing IoT devices, such as wearable sensors, can be applied across a range of diverse applications. Due to the complexity of activity recognition, most wearable activity recognition systems employ multiple homogeneous or heterogeneous sensors to capture excessive information. However, the increased sensor count and the way of multi-channel signal data pose significant challenges to tasks related to human activity recognition. Determining appropriate sensor channels that strike a balance between computational complexity and recognition accuracy emerges as a pivotal concern. In this part, we extend the sparse group lasso mechanism to address human activity recognition challenges and propose a hybrid attention-based multi-sensor pruning and feature selection deep neural network, abbreviated as (Zhou et al. IEEE Internet of Things J (2022) [1]). This architecture excels in conducting feature selection on the basis of sensor pruning. HAP-DNN comprises three modular and detachable components : the feature compression and reconstruction module for fusing and restoring sensor feature information, the feature weight calculation module for calculating sensor channel weights and feature weights, and the learning module for classification, akin to a filter-based feature selection method. Four public activity recognition datasets are used to verify our proposed architecture, and experimental results show that HAP-DNN achieves the best classification performance with the least number of retained feature channels.

Keywords Deep neural network · Hybrid feature selection · Sparse group lasso mechanism · Human recognition activity · Hybrid attention-based multi-sensor pruning

5.1 Problem Formulation

The Internet of Things (IoT) is a burgeoning technology that aims at linking diverse sensing devices to the Internet, enabling the collection of data generated by these sensors. In recent years, the confluence of artificial intelligence advancements and the widespread adoption of wearable sensor devices has propelled the sensor-based

Y. Zhou et al., *Computational Intelligence for High-Dimensional Machine Learning*, SpringerBriefs in Computer Science, https://doi.org/10.1007/978-981-96-2687-8_5

Human Activity Recognition system (HAR) into the limelight. This system, which amalgamates artificial intelligence and IoT technology, has garnered increasing attention for its pivotal role across various industries, including health monitoring [2], intelligent interactive [3], and security monitering [4]. The HAR system is categorized into two main types: video-based and sensor-based. While video-based recognition systems can achieve high accuracy, they may give rise to privacy concerns and escalate computational complexity due to the video data. The sensor-based recognition system, encompassing anonymous binary sensors and wearable sensors, proves adept at circumventing the aforementioned challenges, rendering it particularly fitting for ubiquitous computing [5]. Anonymous binary sensors [6] offer the advantage of privacy preservation, making them well-suited for long-term activity recognition. However, they may pose challenges in capturing finer activities. Wearable sensors-based systems excel in collecting more effective signal data for subsequent processing.

Presently, multi-sensor fusion approaches [7] are predominantly categorized into three classes: data-level fusion, feature-level fusion, and decision-level fusion. Data-level fusion focuses on the collected signal data and fuse similar data based on the sensor type. Feature-level fusion involves extracting the most representative feature vectors from basic data. In body sensor networks (BSNs), feature-level fusion algorithms [8] find extensive use due to their ability to diminish task calculation complexity and minimize communication waste. Building upon the foundation of feature-level fusion, decision-level fusion integrates information from multiple sensors and formulates decisions based on the actual requirements of the application. It needs robust integration between software and hardware, particularly corresponding algorithms. In the realm of HAR, Meng et al. [9] delve into the impact of a single sensor's location on HAR performance. Phan et al. [10] employ a decision tree model to identify spurious classifications, eliminating labels prone to generating inaccurate inferences and thereby enhancing accuracy. However, both approaches are individual model-based methods with limited applicability. In contrast, Cao et al. introduced a multi-sensor fusion and ensemble pruning system (MSF-EP) [11], incorporating diverse pruning metrics to enhance recognition performance. Subsequently, they proposed [12] an adaptive ensemble pruning algorithm recommendation mechanism, building upon MSF-EP, tailored for different subjects. These ensemble approaches typically leverage multiple classifiers for sensors and optimize them through pruning, effectively reducing computational costs while maintaining classification accuracy.

In HAR tasks, establishing the association between specific sensor data and corresponding activities poses a significant challenge. Many sensors encompass multiple channels, and for a particular action, only certain highly correlated channels from sensors at diverse locations may contribute meaningfully. Furthermore, in scenarios where there are very few sensors in practical HAR tasks, mere sensor pruning can have a substantial impact on recognition accuracy. Consequently, we address this issue by implementing channel selection for various sensors, constituting the multi-sensor pruning challenges addressed in our work.

Furthermore, the majority of ensemble methods concentrate on sensor pruning to mitigate computational costs, often overlooking the influence of features within the sensors on HAR tasks. Given the presence of redundant or irrelevant features within sensors, these features not only escalate computational complexity but also contribute to overfitting in classification models. Hence, there is a pressing need to further minimize redundant features by building upon the foundation of sensor pruning.

Integrating l_1 and l_2 penalties, group lasso [13] serves as an extension of lasso [14] specifically designed for group feature selection, finding applications across various domains. However, group lasso primarily emphasizes group-level sparsity and does not ensure intra-group sparsity. Simon et al. [15] introduced a more comprehensive penalty, amalgamating lasso with group lasso, termed sparse group lasso. In the context of wearable human activity recognition tasks, the challenges of multi-sensor pruning and feature selection closely parallel the optimization problems inherent in sparse group lasso regression. Here, features are categorized into multiple groups based on sensor channels, and the optimal performance in activity recognition is attained through both sensor pruning and the selection of features within sensor channels.

Inspired by the sparse group lasso paradigm, we introduce a novel architecture for human activity recognition tasks named HAP-DNN, which leverages hybrid attention-based multi-sensor fusion pruning and feature selection. This innovative network calculates attention weights for both sensor channels and features within channels, thereby reducing the number of sensor channels while selecting optimal features within the retained channels. The key contributions of our work are outlined below:

- We present a deep neural network for multi-sensor fusion pruning and feature selection, integrating an attention mechanism into its architecture.The proposed network comprises three integral components: a feature compression and recon- struction module, a feature weight calculation module, and a learning module.
- Sensor channel attention weight and feature attention weight mechanism: Our approach takes into account both sensor channel redundancy and feature redun- dancy within HAR tasks. This mechanism adeptly transforms the challenges of multi-sensor fusion and ensemble pruning into a unified problem of sensor pruning and feature selection, facilitated by an attention mechanism.

5.2 Hybrid Attention Mechanism

In Fig. 5.1, the comprehensive architecture of HAP-DNN is illustrated. The overall structure comprises three core components: the feature compression and reconstruc- tion module, the feature weight calculation module, and the learning module. The feature compression and reconstruction module facilitate the fusion and restoration of sensor features. Meanwhile, the feature weight calculation module is responsi- ble for determining both sensor channel weights and the weights associated with

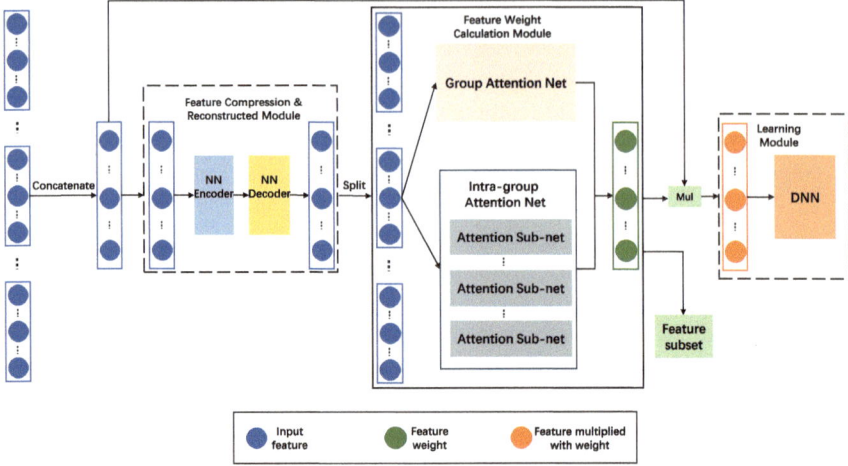

Fig. 5.1 The structure of HAP-DNN. It comprises three main modules: the feature compression & reconstruction module, the feature weight calculation module, and the learning module. The feature weight calculation module, being a pivotal component, is detailed in Figs. 5.2 and 5.3. This module is designed to execute sensor channel pruning and feature selection simultaneously

all features. Sensor channel weights guide the decision on which sensor channels should be retained or discarded. The learning module is dedicated to optimizing the correlation between weighted features and target outcomes. Through the network's training process, adjustments to sensor channel weights and feature weights are made to achieve an appropriate distribution. The architecture exhibits a loosely coupled structure, allowing customization of each module to suit diverse tasks.

5.2.1 Feature Compression & Reconstruction Module

In this module, we employ a straightforward autoencoder for the fusion and recovery of sensor information. The primary objective is to reduce the dimensionality of fused features, thereby facilitating efficient data transmission in an IoT environment. The encoder is constructed with a fully connected layer responsible for compressing sensor feature information. In our implementation, the compression rate is set to 0.1, meaning the output of the encoder is one-tenth of the input. Analogously, the decoder consists of a hidden layer designed to restore the original features from the latent representation. Notably, this module is detachable, offering flexibility for addition or removal based on specific requirements.

5.2.2 Feature Weight Calculation Module

The feature weight calculation module constitutes the nucleus of our proposed network, encompassing both the group attention net and intra-group attention net. Within this module, each channel of the sensor signal is treated as a group, and the features extracted within each channel represent the features within the group. Consequently, all features can be categorized into multiple groups. The group attention net calculates the weights between groups to assess the significance of different sensor channels for the given task. Simultaneously, the intra-group attention net is employed to determine feature weights for each channel. By leveraging these indicators, we effectively reframe the challenges of multi-sensor fusion and pruning into unified problems of sensor pruning and feature selection.

5.2.2.1 Group Attention Net

The architecture of the Group Attention Net is depicted in Fig. 5.2. For each channel of feature vectors, we initially compute both the average and maximum values. Subsequently, we consolidate these average and maximum values from all channels into two vectors, which are then fed into the same multilayer perceptron (MLP) to generate a corresponding vector output. The MLP is structured symmetrically with a hidden layer size equal to half of the input. This design aims to enhance

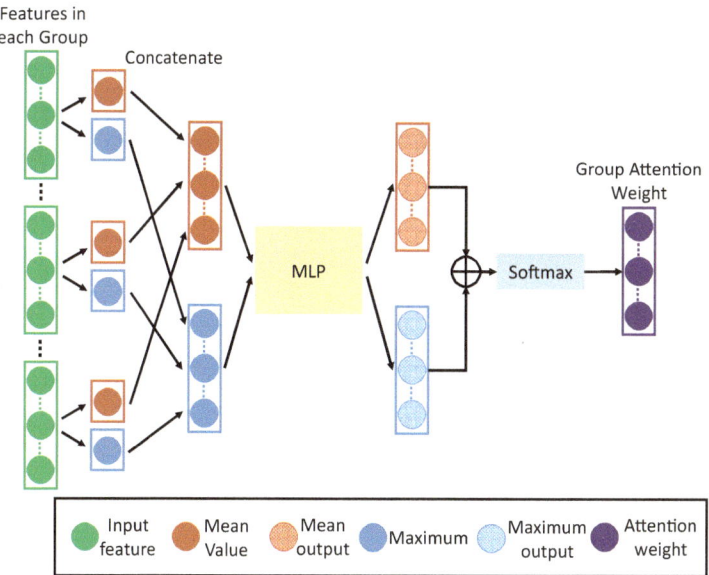

Fig. 5.2 Group attention net

information fusion across different sensor channels and assess the weights of each channel. Following the acquisition of the output, we sum the elements of the two vectors, and apply the softmax activation function to ensure a balanced distribution among channels. These steps can be succinctly expressed by the equation (5.1):

$$G_i = Softmax\left[\Phi\left(v_{avg}(x_i)\right) + \Phi\left(v_{max}(x_i)\right)\right] \tag{5.1}$$

where x_i, G_i are input feature vector and the ith channel weight, v_{avg}, v_{max} are average and maximum value of the ith channel, and $\Phi(\cdot)$ is the MLP operation.

Once the sensor channel weights are determined, they are arranged and selected in descending order. The count of retained sensor channels is contingent on the actual number of selected features. Specifically, the number of selected features is divided by the number of features in each channel, and the result is rounded up to determine the number of retained channels.Furthermore, feature weights are computed in distinct manners. For the sensor channels that are retained, feature weights are calculated using the intra-group attention net. Meanwhile, for the channels that are not selected, their feature weights are directly set to 0.

5.2.2.2 Intra-Group Attention Net

As depicted in Fig. 5.1, the intra-group net comprises multiple attention sub-nets, with an individual attention sub-net designated for each retained sensor channel to compute feature weights within that channel. Drawing inspiration from the non-local network and self-attention Generative Adversarial Network, we have made improvements and simplifications based on these concepts to formulate the attention sub-net. Specifically, considering the sequential nature of sensor signals, we employ 1D convolution with the same size as the feature vector (rather than 1×1 convolution) to transform the input feature vectors into different spaces. The number of convolution kernel channels matches the number of features. The structure of the attention sub-net is illustrated in Fig. 5.3. Feature weights are computed as follows:

$$A_j = v[Softmax(p(x_i)^T q(x_j))g(x_i)] \tag{5.2}$$

here $p(x_i) = W_p x_i$, $q(x_j) = W_q x_j$ and $g(x_i) = W_g x_i$.

W_p, W_q and W_g are 1D convolution weights. $Softmax(p(x_i)^T q(x_j))$ indicates the significance of ith features relative to jth features in each channel. $v(\cdot)$ is 1×1 convolution which is used to adjust the dimensions of output. The output A_j denotes the weight of the jth feature within a sensor channel, rather than representing the final feature weight. It's crucial to consider the weights of different channels to determine the ultimate feature weight.

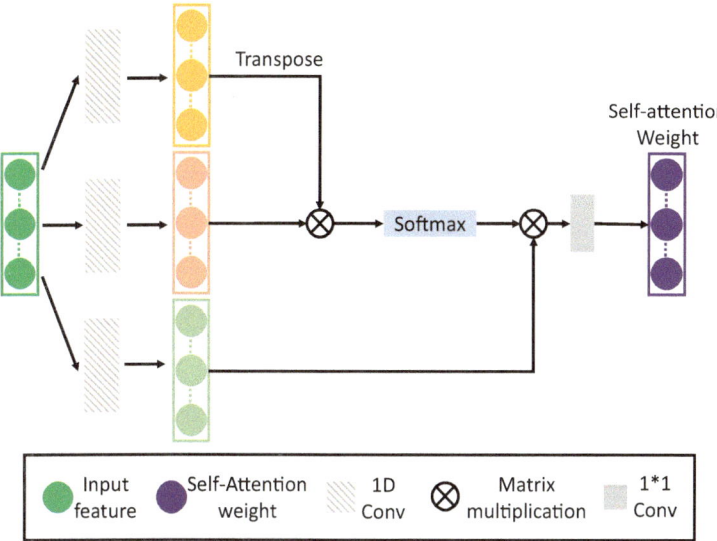

Fig. 5.3 Attention sub-net

5.3 Deep Feature Selection

5.3.1 Learning Module

Within the learning module, we employ a straightforward deep neural network model to optimize the correlation between weighted features and the target. The model primarily consists of a Multilayer Perceptron (MLP) with a hidden layer, where the rectified linear unit (ReLU) activation function is applied after the hidden layer. It is worth noting that the learning module can be substituted with other pre-trained network models to cater to varying task requirements.

Running with back propagation, the learning module keep adjusting sensor channel weights and feature weights to achieve the best distribution. The loss function is:

$$\mathcal{L}(x, \theta) = \arg\min[\frac{1}{N}\sum_{i=1}^{N} \left\| H_{\theta_c}(x_i) - x_i \right\|^2 +$$
$$\frac{1}{N}\sum_{i=1}^{N} Cross\left[L_{\theta_l}\left(G_{\theta_f}(H_{\theta_c}(x_i))\right) - y_i\right] + \lambda R(\theta)] \tag{5.3}$$

where L_{θ_l}, G_{θ_f}, H_{θ_c} are learning module, feature weight calculation module and feature compression&reconstruction module, θ_l, θ_f and θ_c are parameters in these three modules, and $R(\theta)$ is l_2 norm to avoid overfitting of network. λ is a hyper parameter to control the regularization extent. $Cross(\cdot)$ is cross entropy loss function to optimize the network. In our work, we use mean square error loss function and cross entropy loss function for optimization. Training details are shown in section VI.

5.3.2 Calculation Formula of Feature Weight

Each feature's weight is determined by two values: the weight of its channel and the weight within the channel. Thus, for the jth feature in the ith channel, its feature weight is computed as:

$$F_{i,j} = G_i * A_j \tag{5.4}$$

here G_i is the weight of the ith channel, A_j is the jth feature weight in this channel. Regarding the unselected channels, their internal feature weights are set to 0 within the proposed framework. Consequently, the ultimate feature weights for these channels remain 0.

To derive the feature subset, we arrange the final feature weights in descending order and select features with higher weights. Throughout the feature selection process, priority is consistently given to features within the retained sensor channels. Features from discarded sensor channels, having lower corresponding feature weights, are excluded from the selection process. By leveraging sensor channel weights and final feature weights, we can seamlessly execute feature selection while simultaneously pruning sensors.

5.4 Results and Analysis

5.4.1 Datasets

In this section, we use four publicly available multi-sensor wearable activity recognition datasets to validate our network. These datasets include the OPPORTUNITY Activity Recognition Data Set [16] and the Daily and Sports Activities Data Set [17] from the UCI Machine Learning Repository, the Daily Life Activities Data Set [18] from the Machine Learning and Data Analytics Lab, and the Skoda Mini Checkpoint [19]. The sensors used in different datasets are illustrated in Fig. 5.7.

5.4.1.1 Daily Life Activities (DaLiAc) Data Set

The DaLiAc Dataset comprises data from 19 participants engaged in 13 daily life activities. The data were sampled using four SHIMMER sensor nodes with a sampling frequency of 204.8 Hz. Further details about this dataset can be found in [18].

5.4.1.2 OPPORTUNITY Activity Recognition Data Set

The OPPORTUNITY Dataset encompasses daily activity data from 4 subjects, collected from body-worn, ambient, and object sensors. In our work, we exclusively

utilize the data obtained from body-worn sensors, as in [11], focusing on the classification of four activities. The body-worn sensors consist of 7 inertial measurement
units and 12 3D acceleration sensors, providing data on 3D acceleration, 3D rate of
turn, 3D magnetic field, and quaternions.

5.4.1.3 Daily and Sports Activities (DSA) Data Set

This dataset includes data from 19 activities performed by 8 subjects. Sensor units
are calibrated to capture data at a 25 Hz sampling frequency for 5 minutes. The
signals are segmented into 60 segments, each with a length of 5 seconds, resulting
in 480 signal segments for each activity.

5.4.1.4 Skoda Mini Checkpoint (Skoda)

The Skoda Mini Checkpoint dataset encompasses 10 manipulative gestures performed in a car maintenance scenario. Data were sampled using 2 sets of 10 USB
sensors each, placed on the left and right upper and lower arms. The sensor sample
rate is approximately 98 Hz.

5.4.2 Data Segmentation

For the OPPORTUNITY dataset, the raw data is initially segmented using a sliding
window of 500 ms length, with a step size of 250 ms. Given the missing data in this
dataset, we adopt two preprocessing approaches similar to [11]. For sensors with less
than half of the data missing in a segments, we simply repeat the previous value. If
more than half of the sensor data is missing, the entire segment is discarded.

For the DaLiAc dataset, we opt for a sliding window with a length of 5 s and a
step size of 2.5 s for further processing.

For the DSA dataset, since the data have been processed previously, no additional
segmentation is needed.

For the Skoda dataset, we use a sliding window with a length of 2 s and a step
size of 1 s for further processing. Additionally, only the sensors on the left arm are
used for the experiment.

5.4.3 Feature Extraction

For feature extraction in this work, we adopt the minimum, the maximum, the mean
value, variance, skewness, kurtosis, five peaks of the discrete Fourier transform (DFT)

and autocorrelation sequence of each sensor channel as feature construction and extraction criteria as in [11].

We employ distinct feature extraction criteria for different datasets. For the DSA dataset, all the parameters mentioned earlier are used as features. For the OPPOR-TUNITY dataset, the DaLiAc dataset, and the Skoda dataset, we exclude the auto-correlation sequence of each sensor channel. Additionally, min-max normalization is applied to mitigate the range difference between features. Consequently, a total of $21 \times 45 = 945$ features can be obtained for the DSA dataset, OPPORTUNITY dataset has $11 \times 143 = 1573$ features, DaLiAc dataset has $11 \times 24 = 264$ features and $11 \times 60 = 660$ for Skoda dataset.

5.4.4   Experiment Setup

In our research, we employ the repeated random sub-sampling (RRSS) technique, as outlined by Altun et al. [20], for handling the OPPORTUNITY and DaLiAc datasets. Meanwhile, we apply P-fold cross-validation [20] specifically for the DSA dataset. To elaborate, we partition the OPPORTUNITY, DaLiAc, and Skoda datasets into training and testing sets at an 8:2 ratio. Regarding the DSA dataset, we divide all samples into P = 8 partitions, each comprising samples from a distinct subject. Each partition is further divided into training and testing sets at an 8:2 ratio and serves as the validation set iteratively. This cross-validation process is iterated eight times. The training and testing procedures are repeated 10 times, and the resulting average is considered as the conclusive outcome.

In the experiment, we employ three prevalent metrics to evaluate the classification performance: accuracy, weightF, and AUC (area under the ROC curve).

5.4.4.1   Parameter Setting

The model parameters undergo optimization through stochastic gradient descent (SGD) and adaptive moment estimation (Adam), utilizing a batch size of 100. The learning rates are configured at 0.1, and the regularizer weights are set to 0.0001. The model training process is bifurcated into two steps. Initially, we train the feature compression and reconstruction module using the Adam optimizer and the mean square error loss function. Subsequently, the entire network is trained employing the SGD optimizer and the cross-entropy function. The training steps for these two phases are designated as 20 and 40, respectively.

5.4.4.2   Classifier Selection

As our proposed HAP-DNN serves as a filter feature selection method, it exhibits flexibility in combination with various classifiers. Consequently, we assess the

Table 5.1 Average accuracy between four classifiers in four datasets

	NN	SVM	DT	GNB
DSA	**0.9546**	0.9387	0.8793	0.8585
DaLiAc	**0.8826**	0.8589	0.8117	0.7545
OPPORTUNITY	**0.8907**	0.8291	0.8270	0.6460
Skoda	**0.8547**	0.7638	0.7972	0.4938

average classification performance using four distinct classifiers: Neural Network (NN, employing two hidden layers), Gaussian Naïve Bayes (GNB), Support Vector Machine (SVM), and Decision Tree (DT). Given the varying number of features across the four datasets, the range of selected feature numbers, denoted as K, for computing the average accuracy also differs.

As depicted in Table 5.1, NN attains the highest average classification accuracy across the four datasets, with DT and GNB trailing significantly behind NN. Although SVM exhibits comparable accuracy to NN in DSA and DaLiAc datasets, it demonstrates a relatively poor performance in the OPPORTUNITY and Skoda datasets, indicating its lack of stability. Consequently, NN is chosen as the default classifier, and all performance measure comparisons are based on the results obtained with the NN classifier.

5.4.5 Experiment Results

In our experiments, we compare the proposed HAP-DNN with two sensor ensemble pruning algorithms: discriminant pruning (MSF-EP) [11] and dynamic ensemble pruning (META-DEPS) [12]. Additionally, we include convolutional neural network based group lasso (I-CNN) [21] and the traditional sparse group lasso (sgLasso) method using the SLEP toolbox [22] for comparison. Since MSF-EP, META-DEPS, and I-CNN only perform sensor pruning without any feature selection process, we modify these methods to ensure a fair comparison. Specifically, Principal Component Analysis (PCA) is used to reduce the dimensions of features in retained sensor channels after applying these sensor pruning methods, ensuring that the selected feature subsets have the same size. These feature subsets are then used for testing.

Figures 5.4, 5.5, 5.6 and 5.7 display the testing results of all the methods across four datasets. Due to variations in the number of extracted features from dataset to dataset, the range of selected features K also differs. The results indicate that HAP-DNN consistently achieves the best performance across most cases. As ensemble methods, MSF-EP and META-DEPS perform worse than HAP-DNN, and their performance exhibits less stability. As the number of selected features increases, they show slight fluctuations in four datasets. META-DEPS even exhibits a decline when K is higher in the OPPORTUNITY dataset, indicating susceptibility to interference with a larger

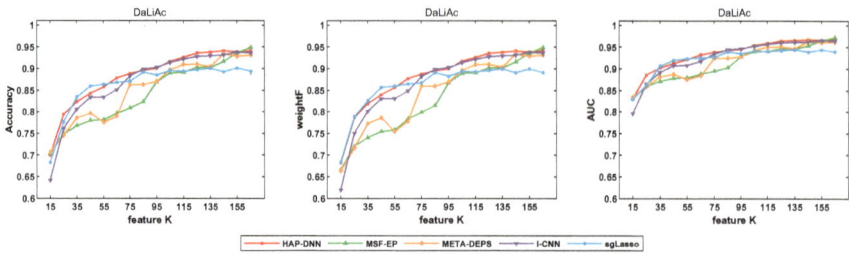

Fig. 5.4 Comparisons among different methods for the DaLiAc datasets. The numbers of selected features are set from 15 to 165

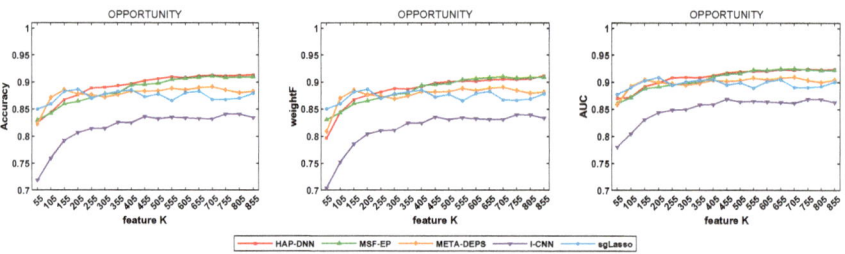

Fig. 5.5 Comparisons among different methods for the OPPORTUNITY datasets. The numbers of selected features are set from 55 to 855

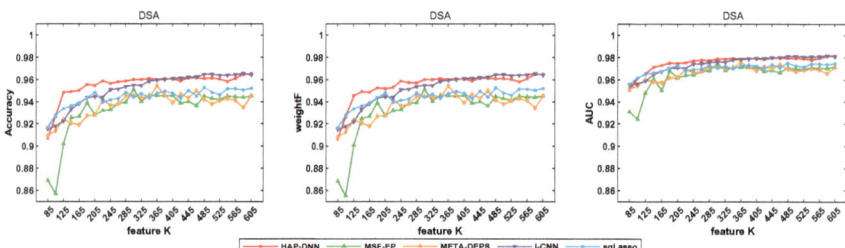

Fig. 5.6 Comparisons among different methods for the DSA datasets. The numbers of selected features are set from 85 to 605

number of features. In contrast, HAP-DNN maintains higher stability across all four datasets, demonstrating its superiority over other methods.

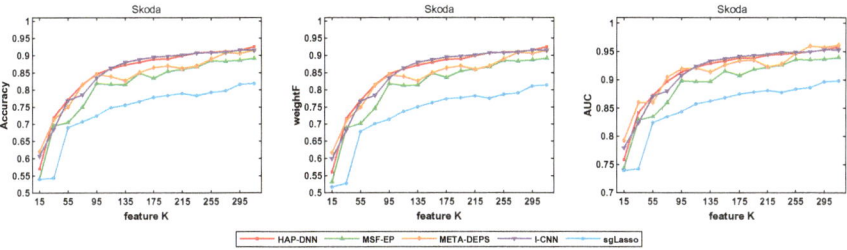

Fig. 5.7 Comparisons among different methods for the Skoda datasets. The numbers of selected features are set from 15 to 315

In comparison to two other lasso-based methods, the performance of I-CNN closely rivals our method on DSA, DaLiAc, and Skoda datasets, underscoring the robust generalization capabilities of neural networks. Notably, I-CNN exhibits the worst performance on the OPPORTUNITY dataset, potentially attributed to the dataset's larger number of sensor channels compared to others, influencing the optimization of the I-CNN network. Conversely, sgLasso, as a conventional sparse group lasso method, fails to attain optimal results across all four datasets. As illustrated in Fig. 5.6, both I-CNN and sgLasso demonstrate lower stability compared to our method.

Additionally, averaging the testing results across the four datasets within the range of feature number K yields comprehensive insights, as presented in Table 5.2. The proposed HAP-DNN outperforms all other methods across the four datasets. In comparison with the ensemble methods, the two lasso-based approaches perform admirably on DSA and DaLiAc, closely approaching the performance of our method. Ensemble methods, on the other hand, demonstrate greater suitability for the OPPORTUNITY dataset than lasso-based methods. In the case of the Skoda dataset, I-CNN proves effective, producing results comparable to our methods, while traditional sgLasso exhibits poor performance.

Table 5.3 presents recognition accuracy using retained sensor channels and all sensor channels in the four datasets. Given the stability in results after a certain number of channels have been utilized, we record the number of retained channels when testing accuracy stabilizes for comparative analysis. The results in Table 5.3 reveal that the accuracy of the proposed HAP-DNN closely aligns with that achieved using all channels across the four datasets. Moreover, our method demands fewer sensor channels than alternative approaches to achieve superior results, highlighting the efficacy of sensor pruning.

Table 5.2 Average testing result for different methods on four datasets

	DSA			DaLiAc			OPPORTUNITY			Skoda		
	Accuracy	WeightF	AUC	Accuracy	WeightF	AUC	Accuracy	WeightF	AUC	Accuracy	WeightF	AUC
HAP-DNN	**0.9556**	**0.9550**	**0.9760**	**0.8826**	**0.8804**	**0.9349**	**0.8924**	**0.8867**	**0.9101**	**0.8480**	**0.8469**	**0.9147**
MSF-EP [11]	0.9329	0.9327	0.9642	0.8416	0.8306	0.9123	0.8866	0.8852	0.9065	0.8082	0.8065	0.8927
META-DEPS [12]	0.9371	0.9369	0.9668	0.8505	0.8421	0.9170	0.8787	0.8769	0.8997	0.8325	0.8317	0.9135
I-CNN [21]	0.9519	0.9518	0.9746	0.8695	0.8667	0.9272	0.8164	0.8137	0.8511	0.8469	0.8459	0.9144
sgLasso [22]	0.9442	0.9441	0.9705	0.8630	0.8617	0.9224	0.8741	0.8740	0.8947	0.7389	0.7303	0.8526

Table 5.3 Accuracy by using the retained sensor channels and all the sensor channels on four datasets

	DSA			DaLiAc			OPPORTUNITY			Skoda		
	Retained	All	Ratio	Retained	All	Ratio	Retained	All	Ratio	Retained	All	Ratio
HAP-DNN	**0.9611**	0.9629	**15/45**	**0.9415**	0.9455	**12/24**	**0.9098**	**0.9115**	**46/143**	**0.9148**	**0.9216**	**22/60**
MSF-EP [11]	0.9452	0.9503	19/45	0.9364	**0.9489**	15/24	0.9086	0.9108	51/143	0.8903	0.9102	28/60
META-DEPS [12]	0.9521	0.9545	18/45	0.9341	0.9386	14/24	0.8897	0.8917	55/143	0.9127	0.9189	28/60
I-CNN [21]	0.9595	**0.9658**	19/45	0.9382	0.9452	13/24	0.8353	0.8417	51/143	0.9084	0.9182	24/60
sgLasso [22]	0.9501	0.9545	21/45	0.9006	0.9060	13/24	0.8830	0.8902	61/143	0.7389	0.7303	27/60

References

1. Zhou, Y., et al.: A hybrid attention-based deep neural network for simultaneous multi-sensor pruning and human activity recognition. IEEE Internet of Things J. **9**(24), 25363–25372 (2022)
2. Oguntala, G.A., et al.: SmartWall: Novel RFID-enabled ambient human activity recognition using machine learning for unobtrusive health monitoring. IEEE Access **7**, 68022–68033 (2019)
3. Liang, J.-M., et al.: Smart interactive education system based on wearable devices. Sensors **19**(15), 3260 (2019)
4. Ordonez, F.J., et al.: In-home activity recognition: Bayesian inference for hidden Markov models. IEEE Pervasive Comput. **13**(3), 67–75 (2014)
5. Qin, Z., et al.: Learning-aided user identification using smartphone sensors for smart homes. IEEE Internet of Things J. **6**(5), 7760–7772 (2019)
6. Gochoo, M., et al.: Unobtrusive activity recognition of elderly people living alone using anonymous binary sensors and DCNN. IEEE J. Biomed. Health Inf. **23**(2), 693–702 (2018)
7. Webber, M., Rojas, R.F.: Human activity recognition with accelerometer and gyroscope: a data fusion approach. IEEE Sens. J. **21**(15), 16979–16989 (2021)
8. Muzammal, M., et al.: A multi-sensor data fusion enabled ensemble approach for medical data from body sensor networks. Inf. Fusion **53**, 155–164 (2020)
9. Meng, L., et al.: Exploration of human activity recognition using a single sensor for stroke survivors and able-bodied people. Sensors **21**(3), 799 (2021)
10. Phan, T.: Improving activity recognition via automatic decision tree pruning. In: Proceedings of the 2014 ACM International Joint Conference on Pervasive and Ubiquitous Computing: Adjunct Publication (2014)
11. Cao, J. et al.: Optimizing multi-sensor deployment via ensemble pruning for wearable activity recognition. Inf. Fusion **41**, 68–79 (2018)
12. Cao, J.: Dynamic ensemble pruning selection using meta-learning for multi-sensor based activity recognition. In: 2019 IEEE SmartWorld, Ubiquitous Intelligence & Computing, Advanced & Trusted Computing, Scalable Computing & Communications, Cloud & Big Data Computing, p. 2019. IEEE, Internet of People and Smart City Innovation (SmartWorld/SCALCOM/UIC/ATC/CBDCom/IOP/SCI) (2019)
13. Yuan, M., Lin, Y.: Model selection and estimation in regression with grouped variables. J. R. Stat. Soc. Ser. B Stat Methodol. **68**(1), 49–67 (2006)
14. Tibshirani, R.: Regression shrinkage and selection via the lasso. J. R. Stat. Soc. Ser. B Stat Methodol. **58**(1), 267–288 (1996)
15. Simon, N., et al.: A sparse-group lasso. J. Comput. Graph. Stat. **22**(2), 231–245 (2013)
16. Chavarriaga, R., et al.: The Opportunity challenge: a benchmark database for on-body sensor-based activity recognition. Pattern Recogn. Lett. **34**(15), 2033–2042 (2013)
17. Barshan, B., Yüksek, M.C.: Recognizing daily and sports activities in two open source machine learning environments using body-worn sensor units. Comput. J. **57**(11), 1649–1667 (2014)
18. Leutheuser, H., Schuldhaus, D., Eskofier, B.M.: Hierarchical, multi-sensor based classification of daily life activities: comparison with state-of-the-art algorithms using a benchmark dataset. PLoS ONE **8**(10), e75196 (2013)
19. Zappi, P., et al.: Activity recognition from on-body sensors: accuracy-power trade-off by dynamic sensor selection. In: Wireless Sensor Networks: 5th European Conference, EWSN 2008, Bologna, Italy, 30 Jan–1 Feb 2008. Proceedings. Springer Berlin Heidelberg (2008)
20. Altun, K., Barshan, B., Tunel, O.: Comparative study on classifying human activities with miniature inertial and magnetic sensors. Pattern Recogn. **43**(10), 3605–3620 (2010)
21. Kim, E.: Interpretable and accurate convolutional neural networks for human activity recognition. IEEE Trans. Ind. Inf. **16**(11), 7190–7198 (2020)
22. Liu, J., Ji, S., Ye, J.: SLEP: Sparse learning with efficient projections. Arizona State Univ. **6**(491), 7 (2009)

Chapter 6
Real-World Case Study

Abstract This chapter explores practical applications of computational intelligence and feature selection across diverse domains. The first section addresses Fault Diagnosis in Industrial Processes, proposing the Improved Localized Feature Selection (LFS) method (Zhou et al. IEEE Trans Instrum Meas (2023), [1], based on Multi-Objective Binary Particle Swarm Optimization (LFS-MOBPSO). This approach optimizes conflicting objectives without resorting to balancing strategies, demonstrating superiority over existing methods in imbalanced scenarios.The second section tackles the Classification of DNA Microarray Data, presenting the Cooperative Coevolutionary Multiobjective Genetic Programming (CC-MOGP) (Qing et al. in Proceedings of the genetic and evolutionary computation conference (2021), [2]) approach. This method transforms multiclass problems, coevolves populations, and employs a cooperative coevolutionary Pareto archived evolution strategy.

Keywords Computational intelligence · Industrial informatic · Practical application · Cooperative coevolutionary multiobjective genetic programming

6.1 Application in Industrial Informatics

6.1.1 Problem Statement

The complexity of modern industrial systems is constantly rising, as are the requirements for the reliability and safety of production processes. Fault diagnosis is an effective method of ensuring the safety of a system and identifies possible anomalies. A variety of fault diagnosis methods have been proposed in the past, which fall into three main categories: model-based methods [3], signal-based methods [4, 5], and knowledge-based methods [6]. Knowledge-based methods, also known as data-driven methods, do not rely on a prior knowledge or precise mathematical models. Instead, artificial intelligence techniques are used to analyze historical data [6]. With the advancement of information technology and the surge in data in the production process, data-driven methods have been widely used.

Y. Zhou et al., *Computational Intelligence for High-Dimensional Machine Learning*, SpringerBriefs in Computer Science, https://doi.org/10.1007/978-981-96-2687-8_6

In real industrial processes, much more data are collected under normal conditions than under fault conditions [7]. Moreover, different faults possibly occur at the simultaneous time, so fault diagnosis can be regarded as an imbalanced multiclassification problem, and it is of great practical significance to study the imbalanced fault diagnosis methods. Data imbalance may bias the algorithm in favor of normal samples, causing faulty samples to be ignored, thus deteriorating the performance of the algorithm. Common imbalance fault diagnosis methods include data-level resampling-based methods [8, 9], which balance the data by upsampling the faulty samples or downsampling the normal samples. As well as algorithm-level cost-sensitive methods [10], which focus the algorithm more on the faulty samples by adjusting the weight sizes of different categories in the cost function. In addition, for the imbalance classification problem, there are also ensemble learning-based methods that combine the advantages of both these methods [11, 12]. Ensemble learning combines the results of multiple independent models to obtain better predictions [13].

In addition, the gathered data often include much redundant information. Feature selection (FS) is a method of selecting a representative feature subset based on certain evaluation criteria, aiming to effectively distinguish between different sample types within the dataset [14]. It is extensively used in the pre-processing phase of fault diagnosis [15–17]. At the same time, class imbalance is often accompanied by problems of imbalance in the high dimensional distribution of feature attributes [18]. As a result, FS for handling imbalanced data is often combined with the above balancing strategies (e.g., resampling and cost-sensitive) to reduce the bias induced by class imbalance [19]. However, the balancing strategy does not always be effective and the order of resampling and feature selection influences the performance of the algorithm [20]. This suggests that using a balancing strategy may lead to the introduction of additional bias, such that the selected feature subset does not represent the true distribution well. Selecting the optimal feature subset for all samples may cause the algorithm to be biased toward features that help distinguish the majority of samples, thereby ignoring the features that are truly important to a small subset of the samples. For example, in a bearing vibration signal dataset, selecting commonly used features such as mean and variance statistics may result in ignoring frequency features, allowing missed or false detection of faults that are enhanced by high or low frequency signals. Therefore, we need to select various feature subsets according to different fault patterns to increase the performance of fault detection. Over the past decade, the property of localization of samples has been explored intensively and applied to enhancing the performance of machine learning models [21–24]. Samples from the same fault pattern are similar, and the similar samples in a localized region are clustered together [25]. At the same time, performing implicit class balancing by partitioning local regions makes it easier to select features that truly represent fault patterns. Inspired by this, instead of combining the balancing strategy with the traditional global feature selection, we consider establishing multiple localized feature selection (LFS) models to deal with the imbalanced fault diagnosis problem.

In recent years, LFS algorithms advanced considerably in the field of data classification. In [26], inspired by the binding of antibody to antigen, each sample is considered as an antigen and an antibody (feature subset) is initialized for each antigen. During cloning, the antigen that retains the most aggregates within a certain range of antibodies with the same label as itself. In [25], maximizing the inter-class distance and minimizing the intra-class distance as a single objective, a clonal selection algorithm is applied to find the optimal feature subset for each region. In [27, 28], inter-class and intra-class distances are considered as two objectives, and the ε-constraint approach is utilized to solve the LFS problem for each formulation. Based on the previous analysis, the essence of LFS is to consider two distances simultaneously to select the optimal feature subset for each local region. However, existing LFS algorithms either convert one of the objectives to a constraint or solve the multi-objective problem by combining the two objectives into a single objective through the weight coefficients. These methods, on the one hand, need to find a suitable set of weight parameters. On the other hand, by fixing an objective, the inability to fully explore the solution space results in affecting the quality of the selected feature subset. Recently, multiobjective evolutionary algorithms (MOEAS) based on GA [29] or PSO [30], have shown their well performance in solving multiobjective optimization problems (MOPs). Unlike the above methods that generate one solution on a single run, this method obtains a set of Pareto solutions through solution interaction. In addition, the derivative-free characterization of MOEAs can better approximate complex Pareto fronts so as to maintain high-quality solutions. Therefore, it is necessary to develop an improved LFS algorithm to enable accurate fault diagnosis.

Based on the preceding discussion, we utilized the spirit of local feature selection to solve the imbalance classification problem in fault diagnosis, and proposed an improved LFS algorithm based on particle swarm optimization (PSO) to obtain the near optimal feature subsets. Specifically, imbalanced multifault diagnosis is modeled as a set of LFS problems, the sample space is divided into local regions centered on each sample, a binary classifier is constructed for each region, and then the final classification result is acquired by voting. In order to maximize the accuracy of the binary classifier, a multi-objective binary particle swarm optimization (MOBPSO) algorithm is applied to solve the LFS problem directly, and it applies a new strategy to select the final solution from the Pareto solution set. This algorithm is named LFS-MOBPSO. Simulation results on Tennessee Eastman Benchmarks and real-world case studies demonstrate that LFS-MOBPSO outperforms existing LFS and imbalanced ensemble classifiers for various types of imbalance ratio settings. Finally, the robust scalability and generalization ability of LFS-MOBPSO is proved by the comparison of LFS-MOBPSO with existing LFS methods on 12 UCI datasets. The contributions of our work can be concluded as follows:

(1) The imbalanced classification problem in fault diagnosis is tackled from a novel perspective by modeling the problem as a set of local feature selection (LFS) problems. Compared with the balancing strategy and global feature selection, LFS better utilizes the advantage of local distribution of data.

(2) MOBPSO, a derivative-free algorithm, is developed to efficiently tackle formulated multi-objective LFS problems directly and obtain near optimal feature subsets, where a new Pareto solution selection strategy is proposed.

(3) Experimental comparison datasets from multiple benchmarks show that our approach has exhibited strong generalization capabilities as well as scalability, and can handle different dimensions of imbalanced classification efficiently.

6.1.2 Methods and Algorithms

6.1.2.1 Problem Modeling

(1) Model overview

Suppose $\{x_i, y_i\}_{i=1}^N$ represents a collection of N training samples, where each $x_i \in \mathbb{R}^M$ denotes the i-th training sample with M-dimensional features, and $y_i \in \{1, \ldots, c\}$ represents its class label, with c being the total number of classes. We suppose that a region exists around the point x_i that maximizes the representation of the original local data and reduces superfluous information simultaneously. This optimal local region denoted as S_i^*, which satisfies:

$$S_i^* = \arg\max_{S_i} H(S_i) \tag{6.1}$$

In the context of simplifying the local region S_i to a hypersphere, it is defined by x_i, f_i, and r_i. Here, $f_i \in \{0, 1\}^M$ represents the feature subset, while $r_i \in \mathbb{R}$ indicates the radius size. The $H(\cdot)$ metric quantifies the quality of the local region S_i and can be tailored to specific requirements. For our fault diagnosis classification needs, we set it to Eq. (6.7). Within S_i^*, we can establish a binary classifier to determine whether a query sample x_q belongs to label y_i. Subsequently, the outcomes of the N regions $\{S_i^*\}_{i=1}^N$ are aggregated to diagnose x_q.

The following are the advantages of this model:

- It optimizes the representation of feature subsets within local regions without the use of any balancing strategy.
- The capability of a local feature subset to represent its corresponding local region can be measured and guarantees the performance of the binary classifier.
- The final result is synthesized from all local regions and works in the role of an ensemble algorithm.

Figure 6.1 illustrates the comparison between global FS and FS. Global FS tackles imbalanced data through a balancing strategy. However, it is necessary to consider which balancing strategy to choose and the order in which the balancing strategy and feature selection are utilized. In contrast, LFS learns the local distribution around each sample and ensures that the central class is not underweighted, thus completing the treatment of imbalance implicitly.

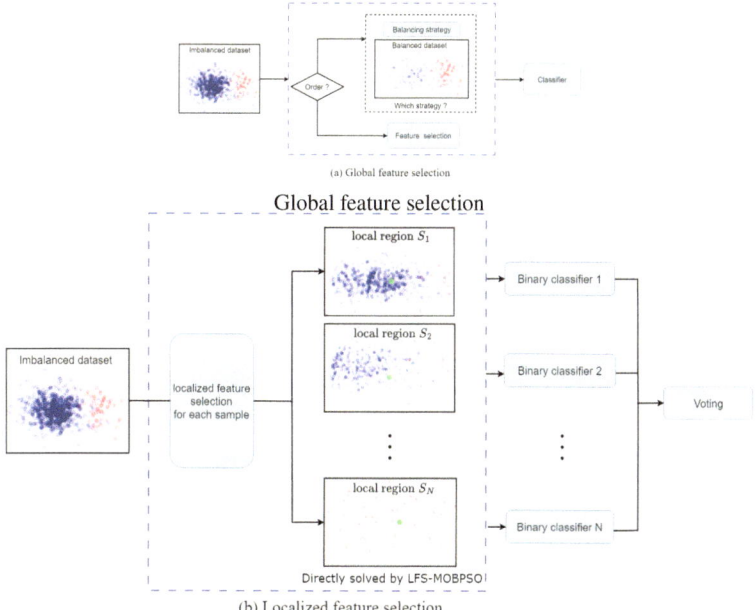

(a) Global feature selection

(b) Localized feature selection

Fig. 6.1 The imbalanced dataset in the graph has two classes, indicated by blue and red dots respectively. The green dots in the right subplot represent the central sample corresponding to each local region

The pseudo code of the model is depicted in Algorithm 4. The two primary issues to address are obtaining $\{S_i^*\}_{i=1}^N$ and constructing the classifier. A detailed solution will be provided in the subsequent section.

Input: $D = \{x_i, y_i\}_{i=1}^N$, τ, α, γ
Output: $\{S_i^*\}_{i=1}^N$
Set $f_i^* = (0, \ldots, 0)^T$, $i = 1, \ldots, N$;
for $k = 1 \rightarrow \tau$ **do**
$\quad$ $f_i^{pre} = f_i^*$;
$\quad$ **for** $i = 1 \rightarrow N$ **do**
$\quad\quad$ $E \leftarrow$ a set of non-dominated f_i. obtained by MOBPSO by solving (6.2);
$\quad\quad$ Set f_i^* equal to member of E which has the highest $NACC$;
$\quad$ **end**
end

Algorithm 4: LFS-MOBPSO

(2) Detailed solution

It is equivalent to obtaining f_i^* and r_i^* defining S_i^* for the first problem. We suppose that f_i^* should make the samples of the same class label as y_i as close as possible to x_i in the corresponding local region, and make the samples of different class labels as far away from x_i as possible. Therefore, f_i^* should satisfy the following conditions:

$$\min_{f_i} \sum_{j:y_j=y_i} w_{ij} \left\| (x_i - x_j) \otimes f_i \right\|_2$$
$$\max_{f_i} \sum_{j:y_j\neq y_i} w_{ij} \left\| (x_i - x_j) \otimes f_i \right\|_2 \qquad (6.2)$$
$$s.t.\, f_i \subset \{0, 1\}^M$$

Here, $\otimes$ represents the element-wise product, $\| \cdot \|_p$ computes the Euclidean distance, and w_{ij} denotes the weight of the distance between x_i and x_j. Through the minimization of intra-class distances and the maximization of inter-class distances within a local region, Eq. (6.2) endeavors to amplify the distinction between samples sharing the same class as x_i and those from different classes.

In order to focus the optimization around the center point x_i, w_{ij} should be proportional to the distance inversely. However, the computation of the distance relies on the coordinate system and there is no prior knowledge of the optimal feature subset to us. Therefore, we use the iterative approach proposed in [27], where w_{ij} is computed as follows:

$$w_{ij} = \frac{1}{N} \left(\sum_{k=1}^{N} \exp\left(-(d_{ij|k} - d_{ij|k}^{\min}) \right) \right) \qquad (6.3)$$

$$d_{ij|k} = \left\| (x_i - x_j) \otimes f_k^{pre} \right\|_2 \qquad (6.4)$$

$$d_{ij|k}^{\min} = \begin{cases} \min_{v:y_v=y_j} d_{iv|k}, & \text{if } y_j = y_i \\ \min_{v:y_v\neq y_j} d_{iv|k}, & \text{if } y_j \neq y_i \end{cases} \qquad (6.5)$$

In this case, f_k^{pre} represents the optimal feature subset of x_k identified in the prior iteration. Initially, f_i^* is initialized as a zero vector, indicating that all distances are assigned equal weights. In the following iterations, w_{ij} is computed using the distances in all regions, as empirical evidence suggests that if two samples are neighbors in one region, they are likely to be neighbors in other regions as well.

Clearly, the issue in model (6.2) constitutes a multi-objective optimization problem (MOP). Typically, there is no single optimal solution for an MOP. All potential solutions are referred to as Pareto solutions. Assuming all optimization problems $\{f_i(x)\}_{i=1}^n$ are to be minimized, we denote x dominates y noted as $x \prec y$, if $\forall i \in 1, \ldots, n, f_i(x) \leq f_i(y)$ and $f(x) \neq f(y)$. A solution that is not dominated by any other solution is termed a Pareto optimal solution. If the solutions within a given set are all Pareto optimal solutions, it is referred to as the Pareto optimal set.

Armanfard et al. [27] utilized the ε-constraint method to convert the inter-class distance into a constraint that is no less than ϵ. By incrementally adding a small value to ϵ until an upper bound on the intra-class distance is attained, the whole Pareto optimal set can be gradually mapped out. However, this method can not approximate the complex Pareto front very well and may result in failure to fully explore the solution space. Toward this end, we use MOBPSO to tackle the problem straightforwardly, as described in the Sect. 6.1.2.2.

Upon gaining the Pareto optimal set of Eq. (6.2), we need to select the final solution from it, which is accomplished in two steps. The first step is to determine the radius of the local region. In the case of region S_i, for example, samples of the same class as x_i are called positive samples, those of different classes are called negative samples. In the space defined by a specific feature subset f_i, the radius value r_i is initially set to 0, and then progressively increased until region S_i meets the following condition:

$$\frac{N_{S_i}^-}{N^-} > \gamma \cdot \frac{N_{S_i}^+}{N^+} \tag{6.6}$$

where γ represents the predefined impurity level, and N^+ and N^- denote the number of positive and negative samples in the complete sample space, respectively. $N_{S_i}^+$ and $N_{S_i}^-$ indicate the number of positive and negative samples within S_i. Here, r_i stands for the radius of the local region determined by f_i. The subsequent step is to select the ultimate solution f_i^*. A normalized accuracy (NACC) has been developed to evaluate the quality of a local area, which is computed as follows:

$$NACC = \frac{1}{2} \cdot \left(\frac{TP_{S_i}}{P_e} - \frac{FP_{S_i}}{N_e} + 1 \right) \tag{6.7}$$

where TP_{S_i} represents the number of samples within S_i that are correctly classified as the positive class. Similarly, FP_{S_i} denotes the number of samples within S_i that are misclassified as the positive class. P_e and T_e represent the number of positive and negative samples in the entire dataset, respectively. When all positive samples are situated in S_i and each sample in the local region has the same label as its nearest neighbor, $NACC$ is equal to 1. Conversely, when each sample in the local region does not have the same label as its nearest neighbor, $NACC$ equals 0.

The typical selection approach involves creating a single objective function by combining multiple objective functions linearly and assigning a weight to each one. Subsequently, the final solution is determined by selecting the one with the highest weighted sum from the solution set. However, as illustrated in Fig. 6.2, if the selection is solely based on the two distance objectives in Eq. (6.2), it could result in choosing the feature subset associated with the left region. This region exhibits a higher concentration of positive samples but also contains incorrectly clustered samples. In contrast, utilizing NACC as the selection criterion tends to prioritize solutions

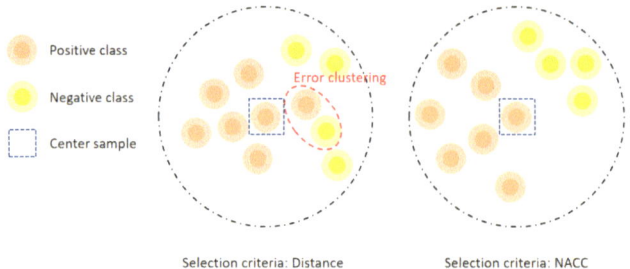

Positive class

Negative class

Center sample

Error clustering

Selection criteria: Distance   Selection criteria: NACC

Fig. 6.2 Local regions obtained based on different solution selection criteria

corresponding to the right region. Despite the relatively sparse samples in this area, the solution is chosen because all samples within it are correctly clustered.

For the second problem, a voting strategy is used to obtain the final diagnosis. Each region S_i^* that is obtained after the previous steps has many positive samples clustered inside it, and these samples are normally in the majority. To distinguish between positive and negative classes in S_i^* is easy. At the same time, given the inherent similarity among real-world data, it is probable that the query sample x_q belongs to regions with positive classes similar to those of x_q, rather than dissimilar ones. Consequently, we can derive the final predicted label $\tilde{y}_q$ for x_q using the following approach:

$$\tilde{y}_q = \arg\max_k \frac{1}{n_k} \sum_{i;y_i=k} V(x_q, S_i^*), \ k \in \{1, ..., c\} \tag{6.8}$$

$$V(x_q, S_i^*) = \begin{cases} 1, & \text{if } x_q \in S_i^* \text{ and } y_q^i = y_i \\ 0, & \text{otherwise} \end{cases} \tag{6.9}$$

where y_q^i represents the classification result of x_q in S_i^*, and n_k indicates the number of samples with label k in the entire training set.

6.1.2.2   Multi-objective Binary Particle Swarm Optimization

Evolutionary algorithms have inherent parallelism and powerful search capabilities, and have made progress in MOPs problems [31, 32]. Among them, Particle Swarm Optimization (PSO) is widely used in FS problems due to its simple implementation, no need for complex parameter settings and operations, and fast convergence speed [33, 34]. PSO was initially developed by Kennedy and Eberhart [35] to simulate the process of birds searching for food through a swarm of particles. While PSO is designed to address optimization problems in real space, Kennedy and Eberhart later introduced Binary Particle Swarm Optimization (BPSO) [36] as an extended

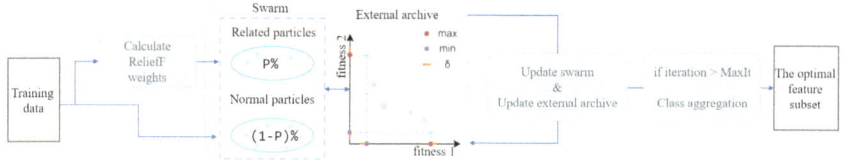

Fig. 6.3 Overview of the proposed MOBPSO algorithm

version for application in binary space. In this study, MOBPSO is utilized to solve model (6.2). The overall process is illustrated in Fig. 6.3.

(1) Variable Definition

During iteration t, the i-th particle in the swarm maintains a position denoted as p_t^i, a velocity denoted as v_t^i, and a record denoted as **pbest**^{i}, which indicates the best position it has searched. The position represents a selection scheme for the feature subset, wherein if the i-th value in the position is 1, the i-th feature will be selected; otherwise, it will not. **gbest** denotes the optimal solution for the population. All these variables are M-dimensional vectors, where M represents the number of decision variables.

(2) Initialization strategy

Since the two objectives in model (6.2) are related to the number of features in a highly linear manner, the solution is difficult to converge. Therefore, α features are randomly selected for each initial particle, which will be deposited in an external archive and guide subsequent iterations. The initialization method outlined in [30] is utilized to enhance search efficiency. This method involves initially employing the ReliefF algorithm [37] to compute the weight of each feature and subsequently normalizing these weights to fall within the range of [0, 1]. A weight closer to 1 signifies a higher importance of the respective feature. Subsequently, all features with weights surpassing a predefined threshold are included in a collection referred to as the relevant set. The particles to be initialized are divided into two parts. One part constitutes $p\%$ of the total number. For each particle in this subset, alpha features are randomly selected from the relevant set and set the corresponding values in the particle positions to 1, while the rest of the values in the particle position are set to 0. On the other hand, the other subset of particles selects alpha features from the entire feature set and assigns the corresponding values in the particle's position to 1, while setting the remaining values in the particle's position to 0. Finally, the velocities of particles are uniformly randomized within the range of $[velmin, velmax]^M$, and $pbest$ is initialized to be the same as the initial position.

Since MOP does not have an optimal solution, an external archive is necessary to preserve the obtained non-dominated solutions. We follow the approach in [38] and partition the external repository into grids. As depicted in Fig. 6.3, we divide the interval between $(1 - \delta)fitness_i^{min}$ and $(1 + \delta)fitness_i^{max}$ into σ segments, where δ and σ are predetermined parameters. Thus, the archive is divided into σ^2 grids, where each grid retains a degree of crowding indicated by the number of particles in that grid.

(3) Update strategy

The d-th element of the i-th particle velocity is updated as follows:

$$
\begin{aligned}
v_{t+1,d}^i &= w * v_{t,d}^i + c_1 * r_1 * (pbest_d^i - p_{t,d}^i) \\
&\quad + c_2 * r_2 * (gbest_d - p_{t,d}^i)
\end{aligned}
\tag{6.10}
$$

Here, w represents the inertia weight, while c_1 and c_2 denote acceleration constants. Additionally, r_1 and r_2 are uniformly distributed random values ranging from 0 to 1. To obtain the **gbest**, we first use roulette selection to select a non-empty grid, where for each grid the probability of being selected is computed based on the reciprocal of crowding degree. Then a random particle is obtained from the grid and its position is used as **gbest**. When there is no update in the archive for several times, it is randomly selected from the archive as **gbest**.

The standard BPSO has some difficulties that lead to slower convergence on some optimization problems, so we adopt the practice of [39] to update the d-th element of the i-th particle position as follows:

$$
R_d^i = \frac{2}{1 + e^{-|v_{t+1,d}^i|}} - 1
\tag{6.11}
$$

$$
p_{t+1,d}^i = \begin{cases}
p_{t,d}^i, & rand > R_d^i \\
1, & rand \le R_d^i \text{ and } v_{t+1,d}^i > 0 \\
0, & rand \le R_d^i \text{ and } v_{t+1,d}^i < 0
\end{cases}
\tag{6.12}
$$

Where R_d^i can be interpreted as the likelihood of whether p_d^i moves in the v_d^i direction. Eq. (6.12) guarantees that the particles' positions are unlikely to change when the speed is small, thereby enhancing the particles' local search capability. The fitness of each particle is then recalculated, and **pbest** is updated to p_{t+1}^i if $p_{t+1}^i \prec p_t^i$, left unchanged if $p_t^i \prec p_{t+1}^i$, and randomly updated to one of these otherwise.

During the end of each iteration, the feasible solutions are added to the external archive and the dominant solutions will be dropped. The external archive is then re-gridded and the crowding of each grid is recalculated. If the number of particles in the archive exceeds the limit, a non-empty grid is first selected using a roulette in which the probability of each grid being selected is calculated based on the degree of crowding. Then a particle is randomly removed from that grid. Repeat this process until the number of particles in the archive is less than the limit. By combining the update strategy of the external archive with the selection method for *gbest*, particles will gradually move into an open area, resulting in obtaining uniformly and widely distributed solutions accordingly.

6.1.3   Results and Insights

We conducted a test on both the simulation dataset and the real-world dataset to evaluate the proposed method for diagnosing faults in imbalanced datasets. Finally, we verified the effectiveness of the improvements made to LFS using the UCI dataset.

6.1.3.1   Experimental Setup

(1) Simulation datasets

In the first part, we test the performance of the proposed algorithm on the TE (Tennessee Eastman) benchmark, which is used to simulate actual chemical co-reaction processes. The generated data are time-varying, strongly coupled, and non-linear, and are widely used to test control and troubleshooting models for complex industrial processes. The TE data can be obtained from http://web.mit.edu/braatzgroup/links.html, which contains both training and testing sets. It has one normal class and 21 fault classes, consisting of 52 features [40]. In the training sets, the normal class comprises 500 samples, while each fault class comprises 480 samples. In the testing sets, the normal class comprises 960 samples, and each fault class comprises 800 samples. To reconstruct the imbalanced dataset, we randomly select a subset of samples from each fault type in the training data and combine them with 500 normal samples to create a new imbalanced training dataset. The imbalance ratio (IR) represents the degree of imbalance by comparing the number of normal samples to the number of samples for each fault.

(2) Real-world datasets

In the second part, we utilized bearing fault diagnosis datasets acquired from the bearing experimental bench shown in Fig. 6.4. The motion of the servomotor-driven rotating shaft simulates the diverse speed fluctuations characteristic of an actual rotating machine's operation. Both the test and support bearings provide support for the entire rotating system. Furthermore, a manual transverse force loader set is employed to replicate the applied loads experienced during the operation of the rotating machinery. Four vibration sensors are utilized to observe vibration in two orthogonal directions for both the test and support bearings. In this experiment, as

Table 6.1 Bearing experimental bench datasets

Condition/Dataset	Instances	Features	Fault type
1200 rpm 0 N	400	132	Normal, Inner race, Outer race, Roller
1200 rpm 1000 N	400	132	
600 rpm 0 N	400	132	
600 rpm 1000 N	400	132	

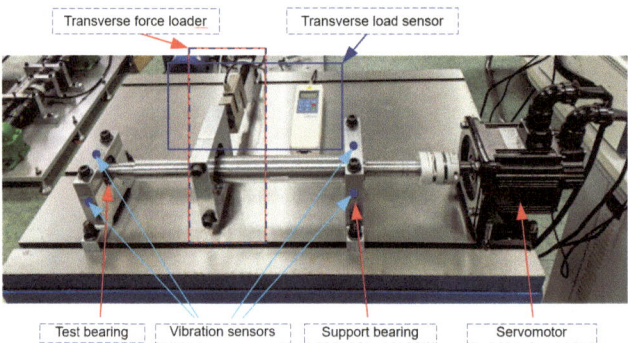

Fig. 6.4 Bearing experimental bench

shown in Table 6.1, two different load conditions (0N and 1000N) and two different speed conditions (600rpm and 1200 rpm) were tested for vibration data for a total of four cases. Each case has four fault types, namely Normal, Inner race, Outer race and Roller. The sampling rate was 25.6 kHz and the acquisition duration was 0.5 s. We set our slicing window to 5 ms, resulting in 100 samples per class and 400 samples per case. Each sensor channel extracts 33 time-frequency domain features (mean, variance, skewness, kurtosis, etc), resulting in a data dimensionality of 132. For each dataset (condition), we utilize 200 samples for training and the remaining samples for testing. In order to obtain an imbalanced dataset, as with the TE benchmark, a portion of each fault class is randomly selected from the training set.

(3) Compared algorithms and Parameter setting

In the first part, we compare the proposed algorithm with PCA [41], Decision Tree [42], Adaboost [43], Bagging [44], Random Forest [45], and LFSDC [27]. Among them, Adaboost, bagging and random forest use decision trees as base classifiers as ensemble algorithms. Additionally, a comparison is conducted with four imbalanced ensemble algorithms using both simulation datasets and real-world datasets. These algorithms include Self-Paced ensemble [46], BalanceCascade [11], SMOTEBoost

Fig. 6.5 The average HV value obtained when the size of the external archive and δ are set to different combined values

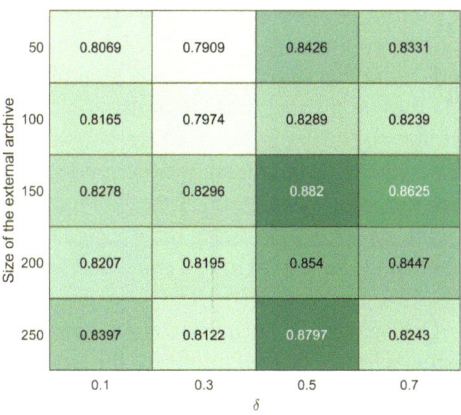

[12] and AsymBoost [47]. Self-Paced ensemble and BalanceCascade utilize sample downsampling, while SMOTEBoost on sample upsampling, and AsymBoost on sample reweighting. Ultimately, our algorithm is compared with two LFS methods: CLONALG-LFS [25] and LFSDC [27]. CLONALG-LFS is an artificial immune system-based LFS method mentioned in the introduction. Additionally, we compare RliefF [48], as it is utilized to generate the particle positions during the initialization phase.

The LFS-MOBPSO is implemented in the MATLAB software. The population size and maximum number of iterations are the basic parameters of the PSO algorithm and are set according to the specific problem. Typically, larger dimensions of the problem necessitate larger population sizes and more iterations. For this study, population sizes of 30 and 100 were assigned to the TE dataset and bearing dataset, respectively. The maximum number of iterations were set at 50 and 70, correspondingly, due to the convergence of populations at these points, as illustrated in Fig. 6.6.

σ is the number of partitions per dimension, which determines the number of grids in the external archive. A too small σ may lead to poor solutions diversity, while a too large σ may result in optimization efficiency problems. We set σ to 30 as suggested

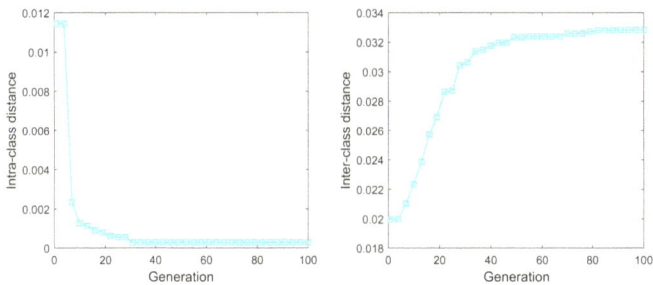

Fig. 6.6 The curve of the changes in each objective value during the iteration

in [38]. In order to determine the remaining parameters of the external archive, the external archive size is set to 50, 100, 150, 200, and 250, and δ is set to 0.1, 0.3, 0.5, and 0.7, in that order. There are a total of 5*4 sets of values, each of which is tested 10 times on the real case. To assess the convergence of the multi-objective solution set, the objectives in Eq. (6.2) are normalized to fall within the range of [0,1], with closer values to 0 indicating better performance. Subsequently, the HV index of the solution set is calculated for these two objectives, using a reference point set at (1, 1). Figure 6.5 demonstrates the average HV values for all combinations. We set the external file size and δ to 150 and 0.5, respectively, because this combination realized the highest average HV.

During population initialization, select features with weights above a threshold for the relevant set, and p controls the number of particles selecting features from this set to guide the direction of updates during iteration. If the threshold and p are set too low, the relevant set may not have enough guiding features or particles, resulting in convergence of the algorithm towards a single update direction. Conversely, setting the threshold and p too high may result in a shortage of common features and particles, which may cause insufficient update momentum in subsequent iterations. Based on the recommendation from [30], we set the threshold and p to 0.52 and 35, respectively. When updating the particles, the velocity limits, referred to as *velmin* and *velmax*, are established as -6 and 6, correspondingly.

For the LFS algorithm, α is the upper bound on the selected feature number. Generally, the testing accuracy increases with the increase of α, but the accuracy change will become stable after exceeding a certain threshold [27]. We set α to half of the problem dimension in order to ensure that α is not too small to affect the performance of the algorithm. τ is the number of optimization iterations for the optimal local feature subset f_i^*. At the outset, given the limited knowledge about f_i^*, it is initialized as a zero vector, assigning equal weight to the distance between all samples. Consequently, τ should be set to a minimum of 2 for Eq. (6.2) to be meaningful. We suggest setting τ to 2 to avoid extra time overhead, since larger τ does not significantly improve performance in practice. γ_{max} max is employed to control the radius of the local region, and we set it to 0.2 as recommended in [27]. To ensure a fair comparison, the aforementioned parameters of LFSDC and the proposed algorithm remain consistent, and the same kNN classifier with k set to 1 is utilized for local regions.

For the other compared algorithms, SVM utilized grid search to adjust parameters, and ensemble algorithms used decision trees as base classifiers with a maximum depth limited to 30 and a quantity of 50. we fine-tuned the range of these compared algorithms on each dataset within the recommended parameters from the corresponding papers, and maintained the best experimental results.

(4) Evaluation Metrics

We primarily evaluate the performance of the algorithm based on classification accuracy. For low-dimensional datasets, classification accuracy refers to the average proportion of correctly identified instances in the test dataset. For high-dimensional datasets, as the data is highly imbalanced, we employ the balanced accuracy calculated as follows:

$$Balanced\ Accuracy = \frac{1}{c}\sum_{i=1}^{c}\frac{TP_i}{N_i} \tag{6.13}$$

where c represents the number of unique labels, N_i stands for the quantity of samples associated with label i, and TP_i indicates the number of samples accurately classified as class i.

In the context of fault diagnosis, we also employ the Macro-F1 score, Macro-Precision, and Macro-Recall [49] for evaluation. These metrics are calculated as follows:

$$Macro - F1 = \frac{1}{c}\sum_{i=1}^{c}\frac{2 * Recall_i * Precision_i}{Recall_i + Precision_i} \tag{6.14}$$

$$Macro - Precision = \frac{1}{c}\sum_{i=1}^{c}Precision_i \tag{6.15}$$

$$Macro - Recall = \frac{1}{c}\sum_{i=1}^{c}Recall_i \tag{6.16}$$

where:

$$Precision_i = \frac{TP_i}{TP_i + FP_i} \tag{6.17}$$

$$Recall_i = \frac{TP_i}{TP_i + FN_i} \tag{6.18}$$

where FP_i represents the number of samples that are incorrectly classified into class i, while FN_i stands for the number of samples that are incorrectly classified as non-class i. Note that these Macro metrics are employed in this context due to the fact that, while the training data is class imbalanced, the test data are balanced (960 for the normal class and 800 for each error class).

6.1.3.2 Comparison on TE Benchmark

Figure 6.7 presents the comparison results at different IRs (IR = 10, 20, ..., 50). PCA has the lowest accuracy as it is a linear feature extraction method that cannot tackle

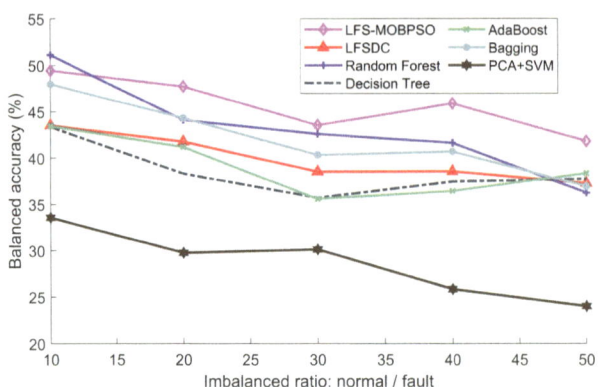

Fig. 6.7 Comparison on TE benchmark at different imbalance ratios (IR = 10, 20, ..., 50)

nonlinear and complex data distributions efficiently when compared to nonlinear model decision tree which has better performance. Adaboost distributes different weights to the samples and has a little better accuracy than the simple decision tree. The bagging algorithm back samples the dataset, constructs multiple weak classifiers, and averages the results as the output with high generalization ability. Random forest builds on the former and performs random selection of features. Both have higher accuracy. LFS-MOBPSO has the highest accuracy and is less affected by imbalance due to the fact that the method facilitates the diagnosis of imbalance faults in two ways. On the one hand, the feature subset is selected based on real data without introducing any bias without using the balancing strategy. On the other hand, the classification accuracy of positive samples is improved by maximizing the difference between positive and negative samples in local regions and increasing the probability of positive samples falling into similar local regions. Therefore, combining the classification results of all local regions improved the final diagnosis accuracy.

Figure 6.8 illustrates the comparison with the four imbalanced ensemble algorithms across various IR values (IR = 10, 20, ..., 100). The results indicate that LFS-MOBPSO demonstrates superior curve trends for all metrics. In particular, its macro-f1 score is notably higher than that of the other methods and is less influenced by data imbalance (the metric value remains relatively stable as the imbalanced ratio increases). In contrast to these imbalanced ensemble algorithms, the proposed method does not employ a balanced strategy for all samples. Our method is equivalent to automatically performing class balancing in local regions. Based on the assumption that samples of the same class are clustered in local regions, the method is able to more accurately reflect the heterogeneous distribution of local data. Additionally, our approach significantly surpasses LFSDC, highlighting the effectiveness of the enhancement.

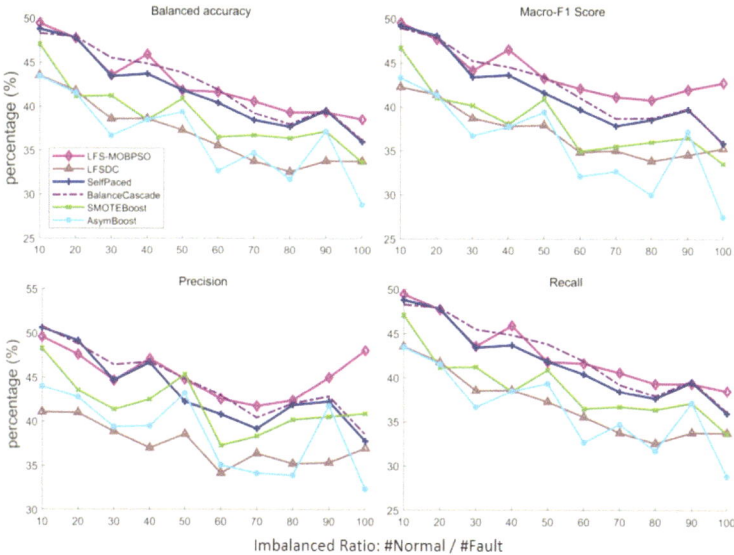

Fig. 6.8 Comparison with four imbalanced ensemble algorithms at different imbalance ratios (IR = 10, 20, ..., 100)

6.1.3.3 Comparison on Real-World Cases

To validate the benefits of the proposed algorithm, experiments were conducted using real-world cases. We experimentally conducted a comparison of the proposed method with LFSDC and the four imbalance integration algorithms for different imbalance ratios (IR = 2, 4, 6, 8, 10) for the real cases presented in Sect. 6.1.3.1. The results are also shown in Fig. 6.9 for comparison with 1200 rpm and 1000 N conditions as an example. Each subplot presents the results obtained for the datasets collected under specific operating conditions. Table 6.2 displays the mean and standard deviation of all metrics resulting from 30 runs of LFS-MOBPSO and other algorithms, each with an IR of 10. The optimal outcomes are highlighted in bold, and we conducted the Wilcoxon significance test [50] at a 5% significance level to compare the performance of the proposed algorithm with that of other methods. Using our method as a reference, "+" means that our method is significantly better than the compared method, "–" means that our method is significantly worse than the compared method, and "=" means that our method is comparable to the compared method. When the IR is 10, the proposed method achieves the best performance under all conditions. This is due to the fact that these imbalanced integration algorithms use diverse advanced balancing strategies, but since the balancing strategies may change the distribution of the dataset and introduce biases, this may result in the model performing worse in the real world than in the training set. In contrast, our approach directly engages with the entire local dataset, allowing it to accurately reflect the local distribution without bias, thereby enhancing the performance and generalization capability of the classifier.

Table 6.2 Comparison and significance testing of all metrics (mean (std)) on bearing fault diagnosis datasets (IR = 10)

Dataset	Metric	SelfPaced ensemble	Balance cascade	SMOTE boost	AsymBoost	LFSDC	LFS-MOBPSO
1200rpm 0N	Balanced accuracy	91.33(0.01) +	91.33(0.01) +	91.33(0.11) +	90.93(0.64) +	98.66(0.01) +	**99.81(0.53)**
	Macro-F1	91.11(0.01) +	91.11(0.01) +	91.06(0.13) +	90.63(0.67) +	98.66(0.01) +	**99.76(0.49)**
	Macro-precision	93.19(0.01) +	93.19(0.01) +	93.16(0.10) +	92.19(0.76) +	98.71(0.01) +	**99.84(0.47)**
	Macro-recall	91.33(0.01) +	91.33(0.01) +	91.33(0.11) +	90.93(0.64) +	98.66(0.01) +	**99.81(0.53)**
1200rpm 1000N	Balanced accuracy	97.06(0.14) +	97.43(0.22) +	95.83(0.17) +	95.66(0.35) +	95.33(0.01) +	**98.67(0.43)**
	Macro-F1	97.07(0.14) +	97.44(0.22) +	95.85(0.18) +	95.69(0.35) +	95.34(0.01) +	**98.67(0.45)**
	Macro-precision	97.15(0.13) +	97.48(0.21) +	95.94(0.17) +	95.89(0.31) +	95.40(0.01) +	**98.73(0.41)**
	Macro-recall	97.06(0.14) +	97.43(0.22) +	95.83(0.17) +	95.66(0.35) +	95.33(0.01) +	**98.67(0.43)**
600rpm 0N	Balanced accuracy	86.04(0.92) =	85.07(1.22) +	84.77(2.22) +	84.74(1.30) +	85.67(0.24) +	**86.33(1.18)**
	Macro-F1	85.70(1.02) =	84.65(1.35) +	84.31(2.41) +	84.27(1.32) +	84.87(0.21) +	**85.74(1.21)**
	Macro-precision	86.69(0.93) +	85.34(1.08) +	85.60(2.15) +	85.76(1.37) +	88.31(0.19) +	**88.65(1.17)**
	Macro-recall	86.04(0.92) =	85.07(1.22) +	84.77(2.22) +	84.74(1.30) +	85.67(0.24) +	**86.33(1.18)**
600rpm 1000N	Balanced accuracy	83.15(0.83) +	83.30(0.35) +	84.34(0.01) +	80.33(0.89) +	86.44(0.23) +	**93.06(1.57)**
	Macro-F1	82.95(0.90) +	83.12(0.36) +	84.15(0.01) +	79.53(0.87) +	86.02(0.21) +	**93.00(1.54)**
	Macro-precision	85.24(0.61) +	85.35(0.22) +	85.73(0.01) +	82.18(0.85) +	87.06(0.19) +	**93.69(1.48)**
	Macro-recall	83.15(0.83) +	83.30(0.35) +	84.34(0.01) +	80.33(0.89) +	86.44(0.23) +	**93.06(1.57)**

6.1.3.4 Impact of the Solution Selection Strategy

In order to verify the impact of using NACC as the selection criterion, we performed experimental comparisons on the datasets presented in Sect. 6.1.3.1. Table 6.3 presents the classification accuracies attained by selecting solutions with weighting method and NACC, respectively, when the IR is 10. Here, we utilize the formula $\lambda * objective_1 + (1 - \lambda) * objective_2$ to assign weights to the two objectives in Eq. 6.2, and subsequently test λ values of 0.1, 0.3, 0.5, 0.7, and 0.9. The optimal λ is observed to vary across different datasets. On the contrary, in most cases, the optimal classification accuracy can be achieved by using NACC as the selection criterion. This is because there are two benefits of using NACC. Firstly, NACC ensures

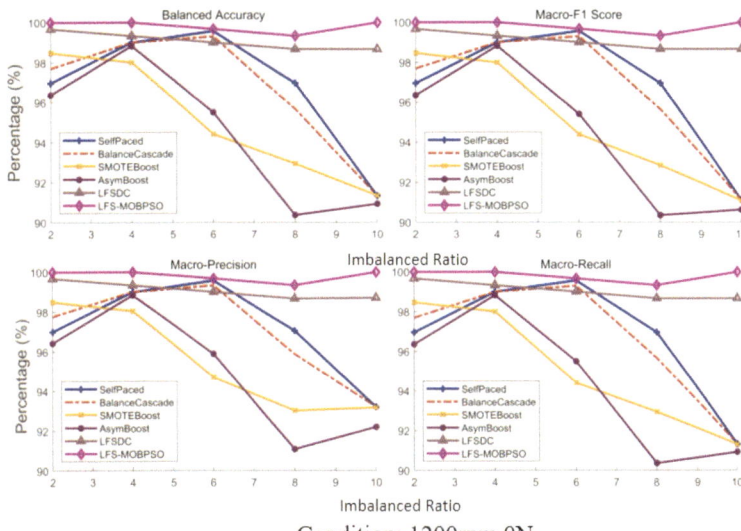

Condition: 1200rpm 0N.

Fig. 6.9 Comparison with LFSDC and four imbalanced ensemble algorithms on real-world cases

Table 6.3 The classification accuracies (%) obtained using different solution selection criteria when IR is 10

Dataset	Weight					NACC
	0.1	0.3	0.5	0.7	0.9	
600rpm 0N	78.21	78.33	**84.01**	83.33	78.05	82.33
600rpm 1000N	85.35	86.02	90.01	91.69	89.77	**93.36**
1200rpm 0N	99.28	99.08	99.00	98.33	95.20	**99.83**
1200rpm 1000N	98.11	98.05	98.00	96.33	88.00	**98.67**
AVG	90.24	90.37	92.76	92.42	87.76	**93.55**

that there are as many positive samples in the local region as possible, thus enhancing the probability of positive test samples falling into that region. Secondly, it can avoid the error clustering phenomenon illustrated in Fig. 6.2, thereby increasing the classification accuracy of the test samples in the local region.

Table 6.4   The impact of sample size and feature number on computation time (seconds)

#Features	#Samples			
	10	100	500	1000
10	1.874	15.24	137.5	383.2
100	2.525	17.23	144.4	344.2
500	2.569	19.6	186	797.1
1000	2.998	19.91	250.7	1314

6.1.3.5   Discussion

(1) Computational time

Sample size and feature number are usually the main factors affecting computation time. We randomly generate simulated datasets with different sample sizes and feature numbers in order to evaluate the computational time of the proposed algorithm. Table 6.4 displays the computational time of LFS-MOBPSO on these datasets, using a population size of 30 and the number of iterations is 50. It is evident that for small sample sizes, the computation time shows little sensitivity to the number of features. The sample size is the main factor that affects the computational time as LFS is a multimodel algorithm, and the number of models increases with the increase of sample size. Therefore, on datasets with a large number of samples, the computation time is relatively long. However, it is worth noting that the optimization process is independent among different models, so this problem can be mitigated by parallel computing.

(2) Strengths and limitations

Through learning the characteristic representation of real local distribution, LFS-MOBPSO can effectively deal with imbalanced data. It also shows strong generality and can be applied directly to different types of datasets. However, the algorithm still has some limitations. Firstly, the LFS framework that is used to partition the local regions based on each sample is time-consuming when dealing with large datasets. Secondly, employing heuristic algorithms to optimize objectives enhances overall performance by circumventing approximations to the objective function, but it leads to a decrease in stability, as evidenced by the standard deviation in Table 6.2.

6.2   Application in Bioinformatics

6.2.1   Problem Statement

In the field of biology, microarray technology [51] has made it possible to measure the expression levels of 30,000-40,000 human genes in a single experiment. The vast amount of DNA microarray data generated by this technology is highly significant for medical diagnosis and genetic analysis [52]. However, the classification of microarray data itself poses a highly challenging task due to the large sample size and the presence of many imbalanced multi-class data.

Feature selection plays a crucial role in reducing the dimensionality of high-dimensional data, as it can eliminate irrelevant and redundant features from the original data. Current feature selection methods are generally classified into three categories: filters, wrappers, and embedded methods [53]. Filters typically directly return a subset of features from the original feature space for selection. Wrappers employ a learning model to evaluate the goodness of feature subsets [54], providing more accurate feature selection compared to filters. Embedded methods strike a balance between the two.

For microarray datasets, traditional feature selection methods are limited by the constraint of sample size, making it challenging to obtain sufficient statistical information. Therefore, further optimization is needed on traditional feature selection methods. In recent years, feature selection methods based on evolutionary algorithms have gained increasing attention. Evolutionary algorithms are widely applied across various domains due to their strong adaptability and high precision. Additionally, their relatively simple framework significantly reduces the computational cost of the algorithm.

In this chapter, we will introduce one of our studies, the CC-MOGP method. Firstly, genetic programming (GP) using a tree-based representation is used for embedded FS. Secondly, we proposed a cooperative coevolutionary Pareto archived evolution strategy (CC-PAES) as the environmental selection strategy to minimize the three objectives that are suggested in [55]. Finally, we decompose a multiclass problem into a set of treatable binary problems and coevolve the corresponding populations for each problem.It converts a multiclass problem into a set of tractable binary problems and coevolves the corresponding population. And a cooperative coevolutionary Pareto archived evolution strategy (CC-PAES) is employed to approximate the Pareto front.

6.2.2   Methods and Algorithms

To handle the high dimension problem, our method employs the FS method proposed in [55] to enhance the intrinsic FS of GP. And the multiclass problem is decomposed into corresponding binary class problems that regard the samples of certain class as positive samples while all other samples as negative. Each binary problem is assigned to certain population and all the populations cooperatively coevolve. Each individual in a population is a GP-tree based discriminant function that works as a binary classifier using zero as its threshold. The adopted evolutionary strategy (in other words, the fitness function and the selection strategy) is CC-PAES that evaluate a new individual based on its domination level compared with the solutions in an external archive and its effect on the synergy score when it was accepted by the archive. Because CC-PAES organizes the population as an archive, the words "population" and "archive" share the same concept in this paper. Besides, the domination level is computed by considering three objectives that need to be minimized: $objective(1)$: false positives (FPs); $objective(2)$: false negatives (FNs); and $objective(3)$: the number of leaf nodes in the tree.

Decreasing training error is a fundamental principle in the theory of machine learning. Only the classification accuracy cannot completely describe the performance of the classifier. For a class imbalance dataset, the prediction of a classifier is biased towards the majority class but also has good performance in the training error. To avoid this, FPs and FNs that evaluate both the accuracies on the majority class and the minority class are taken into regard. Besides, according to the Occam's Razor principles, we use the third objective to restrict the tree size, that is, to control the complexity of classifier. At the same time, this objective can effectively facilitate the FS process.

The schematic representation of the proposed CC-MOGP is given in Fig. 6.10. Our algorithm takes turns to produce a new individual in each population via the mutation and crossover operators proposed in [55] per each generation. CC-PAES is used to decide whether to accept a new individual according to its domination level and effect on the synergy score. The order is that if CC-PAES can't make a decision merely based on the domination level (e.g., the new individual share the same domination level with all of the individuals in the archive), synergy score is used to give its final decision. Before computing the synergy score, all of the decisions are governed by the archive itself. After that, the proposed synergy test

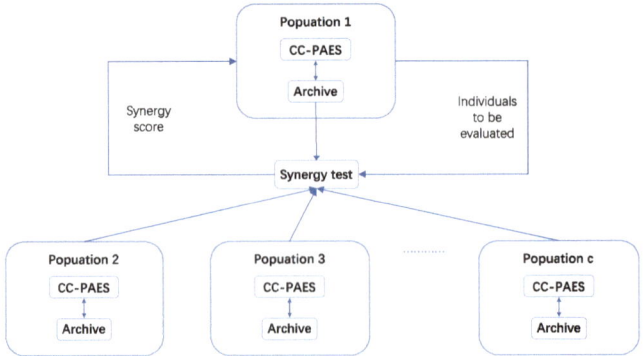

Fig. 6.10 The schematic representation of the proposed CC-MOGP algorithm

method integrates all the archives to construct a complete solution and evaluates the difference of the archive in synergy score after accepting the new individual. If the score increases after replacing an individual of archive with the new individual, then, keep the replacement. The details of CC-PAES and synergy tests are shown in the following subsections and the pseudocode of CC-MOGP is given in Algorithm 5. In this pseudocode, $\mathbf{A}_j$ indicates the jth archive that associated with the jth class. Line 4 is performed based on the classification accuracy proportionate.

Input: Initialization: $S = \{\mathbf{A}_j | j \in N^+, j \leq c\}$ [56].
while $Evaluation^{Current} \leq Evaluation^{Maximum}$ **do**
    for $\mathbf{A}_j \in S$ **do**
        Select a current individual x from $\mathbf{A}_j$.
        Generate a new individual y through mutation and
            crossover on x.
        $\mathbf{A}_j = \text{CC-PAES}(x, y, j)$.
    end
end

Algorithm 5: CC-MOGP

6.2.2.1 CC-PAES

Diversity of population is a fundamental issue for the evolutionary algorithm (EAs) especially for our method that employs an ensemble technique on the archive. In traditional fashion, genotype (infix expressions of the discriminant functions) is a straightforward strategy to measure diversity. Only the individual with different genotypes can be accepted by the archive. However, different genotypes may have the same phenotype (the same response to the training samples) while only the phenotype is directly related to the classification performance. On this premise, we use phenotype as the measure of diversity and make each phenotype appear only once in the archive. Use *Pred* to indicate the phenotype of an individual. *Pred* is a binary vector. $Pred_i$ is its ith variable. $V(\mathbf{o}_i)$ is the output value of the solution for the ith data point $\mathbf{o}_i$, then

$$Pred_i = \begin{cases} 1, \text{if } V(\mathbf{o}_i) > 0 \\ 0, \text{else} \end{cases} \tag{6.19}$$

$Pred_i = 1$ illustrates that the individual believes $\mathbf{o}_i$ belongs to its positive class and vice versa. This is what we call the response of individual.

The pseudocode of CC-PAES is outlined in Algorithm 6. As it is shown, firstly, CC-PAES compares the *Pred* of y (donated by $y.Pred$) with that of the solution in the archive. If it is same as any solution in the archive, retain the one with less number of leaf nodes (the third objective). Otherwise, CC-PAES evaluates the domination between offspring y and its parent x. If x is dominated by y, replace x with y in the archive. If they don't dominate each other, compute the synergy score of the original archive ($A_j.synergy\_score$) and that of the archive whose solution x is replaced by y ($A'_j.synergy\_score$). Then, return the archive with greater $synergy\_score$. The synergy test method is described in the next subsection.

Input: x, y, j.
Output: A_j.
if $\exists\, s \in A_j, \ s.Pred == y.Pred$ **then**
    if $y.objective(3) < s.objective(3)$ **then**
        | $A_j =$ replace s with y.
    end
else
    if $y \prec x$ **then**
        | $A'_j =$ replace x with y.
    end
    else if $x \nprec y$ **then**
        $A'_j =$ replace x with y.
        $A'_j.synergy\_score =$ synergy test(A'_j)
        $A_j.synergy\_score =$ synergy test(A_j)
        if $A'_j.synergy\_score > A_j.synergy\_score$ **then**
           | $A_j = A'_j$
        end
    end
end

Algorithm 6: CC-PAES

6.2.2.2 Synergy Test

Suppose $A_j = \{a_{1j}, a_{2j}, ..., a_{nj}\}^T$ denotes the vector consisting of the probability that the ith ($1 \le i \le n$) training sample belongs to class j, where n is the sample size. a_{ij} drawn in [-1, 1] is the probability that o_i belongs to class j. $R = \{A_1, A_2, ..., A_c\}$ is the judgment matrix. And o_i is assigned to class k if $k = \arg\max_j(a_{ij})$. a_{ij} is obtained by combining the judgments of all solutions in the archive using a negative voting schema defined in (6.20) and (6.21). And that is what we call the ensemble technique.

$$a_{ij} = \frac{1}{N}\sum_{n=1}^{N} b_j^n(i) \tag{6.20}$$

$$b_j^n(i) = \begin{cases} +(1.0 - \dfrac{FP_j^n}{FP_j^{max}}), & \text{if } V_j^n(o_i) > 0 \\[4mm] -(1.0 - \dfrac{FN_j^n}{FN_j^{max}}), & \text{otherwise} \end{cases} \qquad (6.21)$$

where $b_j^n(i)$ is the reliability of the judgment (made by the nth solution of the jth archive) that the ith sample belongs to class j. N is the number of solutions in an archive. $V_j^n(o_i)$ is the output value of the nth solution of the jth archive for the ith sample. FP_j^n and FN_j^n, respectively, represent the number of FPs and FNs made by the nth solution of the jth archive on the training data. FP_j^{max} and FN_j^{max} are the maximum possible FPs and FNs for the solutions in the jth archive.

$$\mathbf{R} = \begin{bmatrix} a_{11} & a_{12} & \cdots & a_{1c} \\ a_{21} & a_{22} & \cdots & a_{2c} \\ \vdots & \vdots & \ddots & \vdots \\ a_{n1} & a_{n2} & \cdots & a_{nc} \end{bmatrix} \qquad (6.22)$$

Equation (6.22) gives an example of judgment matrix. Each row represents the judgments given by c archives for a sample, that is, the probabilities that such sample belongs to their associated classes. And a sample is classified into the class whose associated archive has the highest probability value. Each column is the combination of all the judgments of an archive for the training data. It can be inferred from (6.20) and (6.21) that every single change of solution in the archive could affect its judgment vector accordingly and this effect is not only on a single value of the vector but always on the whole vector. Traditional optimization strategies consider each column independently. For a binary problem, suppose the ground truth of the training data is $\{1, 1, 0, 0, 0, 0\}$ where "1" indicates the positive samples. The objective for conventional methods is to achieve a model with the ideal judgment vector $\{+1, +1, -1, -1, -1, -1\}$ where "±1" means that the model predicts the sample into positive (+) or negative (−) class with 100% confidence. Hence, it requires solutions that drive closely toward the ideal vector. On the contrary, based on the mechanism of how the judgment matrix makes a decision, our goal is to make the desired difference in a row sight as great as possible. Suppose the 1th sample belongs to class 1, that is, a_{11} should be positive while $a_{12}, a_{13}, ..., a_{1c}$ desired to be negative. Our objective is to make the difference between a_{11} and $a_{12}, a_{13}, ..., a_{1c}$ as large as possible. That is the primary concept of *synergy* in this paper. And it is defined as follows:

$$(\mathbf{J} - \mathbf{A}_j) \odot (\mathbf{L} - \mathbf{S}_j) = \mathbf{D} = (x_{ij})_{(M \times C)} \qquad (6.23)$$

$$synergy\_score = \sum_{i=1}^{M} \sum_{j=1}^{C} \frac{1}{1 + e^{-2x_{ij}}} \qquad (6.24)$$

where M and C denotes the number of data points and classes respectively. $\mathbf{S}_j = \{l_1, l_2, ..., l_n\}^T$, $l \in \{0, 1\}$ is the ground truth of the jth population. If the $\mathbf{o}_i$ belongs to class j, $l_i = 1$; otherwise, $l_i = 0$. $\mathbf{L} = \{\mathbf{S}_1, \mathbf{S}_2, ..., \mathbf{S}_c\}$. $\mathbf{R}$ is the judgement matrix obtained by combined all the archives' judgment vectors. Similarly, $\mathbf{L}$ is the label matrix that combines all the archives' ground truth. Our purpose is to assess the *synergy_score* of a given archive $\mathbf{A}_j$. For this purpose, we compute the difference between the judgement vector of $\mathbf{A}_j$ and that of other archives by the operation $\mathbf{J} - \mathbf{A}_j$. Accordingly, perform the same operation on the label matrix. Let $\mathbf{R}' = \mathbf{R} - \mathbf{A}_j$ and $\mathbf{L}' = \mathbf{L} - \mathbf{S}_j$. r and l are the elements in these matrixes respectively. It can be inferred that:

(1) If $\mathbf{o}_i$ belongs to class j and it is correctly classified, $r'_{ik} < 0$ $(k \neq j)$.
(2) If $\mathbf{o}_i$ belongs to class j, $l'_{ik} = -1$ $(k \neq j)$
(3) If $\mathbf{o}_i$ belongs to class p $(p \neq j)$ and it is correctly classified, $r'_{ip} > 0$.
(4) If $\mathbf{o}_i$ belongs to class p $(p \neq j)$, $l'_{ip} = 1$.

According to the above inferences, it can be concluded that if a data point $\mathbf{o}_i$ is correctly classified, the ith row of the $\mathbf{D}$ matrix shouldn't have any negative element. In other words, the row with a negative element means it isn't correctly classified. And the greater the absolute value of the negative element is, the more serious the mistake is. To clearly illustrate the above procedure, we have given an example shown in Fig. 6.11. Suppose there are three data points that belong to class 1, class 2, and class 3 respectively. $\mathbf{o}_1$ is falsely assigned to class 3 because 0.7 >0.2 as marked in the red frame. After the above transformation, the difference between 0.7 and 0.2 is presented in a negative number in $\mathbf{D}$. Then, based on the properties of Eq. (6.24), there are small benefits of optimization on the samples that have already been successfully classified (e.g., $\mathbf{o}_2$ and $\mathbf{o}_3$) but great motivation on the optimization of the samples that have not sufficiently classified ($\mathbf{o}_1$).

Fig. 6.11 Part of the process of computing synergy

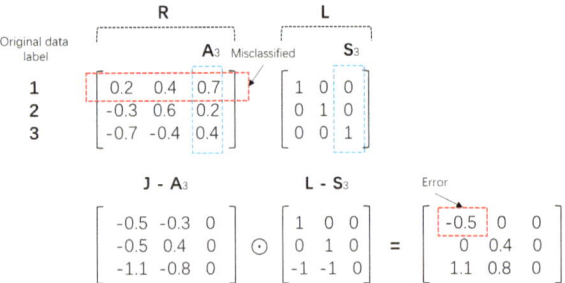

Table 6.5 Datasets

Dataset	Features	Instances	Classes	Smallest class	Largest class
SRBCT	2,308	83	4	13	35
9Tumors	5,726	60	9	3	15
Leukemia1	5,327	72	3	13	53
Brain Tumor1	5,920	90	5	4	67
Leukemia2	11,225	72	3	28	39
Brain Tumor2	10,367	50	4	14	30
Lung Cancer	12,600	203	5	3	68
11Tumors	12,533	174	11	4	16

6.2.3 Results and Insights

6.2.3.1 Experiment Setting

To demonstrate the effectiveness of our method, we have tested it on 8 benchmark data that are publicly available at http://www.gems-system.org. Table 6.5 shows the number of features, the number of samples, the number of classes, and the sample percentages of the largest and smallest classes the ten datasets. As we can see from the table, the number of features of these datasets is much larger than the number of samples, and the gap between the largest and the smallest class is relatively large, so the datasets are highly unbalanced. Therefore, we used the following accuracy formula to persuasively evaluate the performance.

$$Accuracy = \frac{1}{c} \sum_{j=1}^{c} \frac{\text{TP}_j}{|S_j|} \tag{6.25}$$

where c is the number of classes in the dataset, TP_j denotes the number of correctly identified instances in class j, and $|S_j|$ is the sample size of class j. All classes are treated equally with the weight of $1/c$.

To make the experimental results reliable, we repeat tenfold cross-validation for 30 times. The training data is normalised using the Z-score normalisation method. Furthermore, based on the means and the standard deviations of the features of the training data, the test data is also Z-score normalized.

Part of the parameter setting is shown in Table 6.6. These parameters are standard parameters employed in many GP simulations.

6.2.3.2 Comparison of Results and Analysis

To systematically study the performance of the proposed method, we compared CC-MOGP with some state-of-the-art EAs including VLPSO [33], PPSO [57], ASMiGP

Table 6.6 Parameter settings

Parameter	Value
Initial population	Ramped half-and-half [56]
Function evaluations for each archive	400,000
Maximum depth of tree during initialization	6
Maximum allowable depth of tree	10
Probability of crossover	0.8
Probability of mutation	0.2
Maximum archive size	50
Set of function ($\mathcal{F}$)	$+, -, \times, \div$
Set of terminal ($\mathcal{T}$)	Features of dataset, random constant value

Table 6.7 Best and average accuracy of each method

Dataset	Method	Best	Mean /Std	S	Dataset	Method	Best	Mean /Std	S
SRBCT	PPSO	**100.00**	95.78/1.96	−	Leukemia2	PPSO	**100.00**	**96.74/1.64**	=
	VLPSO	**100.00**	**99.67 /0.52**	=		VLPSO	94.44	91.56/1.67	−
	PS-NSGA	98.35	96.35/2.25	−		PS-NSGA	97.22	92.22/3.84	−
	ASMiGP	96.67	92.53/2.13	−		ASMiGP	97.00	91.92/2.66	−
	CC-MOGP	**100.00**	99.64/0.52			CC-MOGP	**100.00**	96.52/1.74	
9Tumors	PPSO	65.00	**59.28/2.08**	+	Brain Tumor2	PPSO	74.58	68.75/4.24	=
	VLPSO	61.67	55.11/4.71	−		VLPSO	73.33	66.78/4.10	=
	PS-NSGA	64.31	58.30/2.79	=		PS-NSGA	78.75	**73.07/6.99**	+
	ASMiGP	53.75	45.76/5.82	−		ASMiGP	72.64	60.75/8.14	−
	CC-MOGP	**67.73**	57.28/3.38			CC-MOGP	**83.33**	67.35/6.87	
Leukemia1	PPSO	95.42	94.37/1.36	−	Lung Cancer	PPSO	84.11	79.38/3.26	−
	VLPSO	97.92	93.31/2.34	−		VLPSO	94.08	89.47/2.18	−
	PS-NSGA	94.00	91.98/3.32	−		PS-NSGA	91.49	88.48/2.58	−
	ASMiGP	96.22	92.74/2.67	−		ASMiGP	94.32	90.72/1.73	−
	CC-MOGP	**99.17**	**97.52/1.46**			CC-MOGP	**97.14**	**94.49/1.09**	
Brain Tumor1	PPSO	**82.08**	74.40/3.67	−	11Tumors	PPSO	83.20	76.83/2.91	−
	VLPSO	79.17	71.19/3.52	−		VLPSO	85.16	80.81/2.32	−
	PS-NSGA	79.72	73.81/4.38	−		PS-NSGA	84.81	83.94/1.84	−
	ASMiGP	76.22	68.23/4.03	−		ASMiGP	88.90	85.64/2.96	−
	CC-MOGP	81.97	**78.90/1.51**			CC-MOGP	**95.77**	**92.37/1.83**	

[55] and PS-NSGA [29]. 30 times of tenfold cross-validation were done for each method. Table 6.7 shows the best and average test accuracy of these methods. The highest best and average accuracy obtained on each dataset are bold. Colum S represents the statistical Wilcoxon significance test results (with 5% significance level) of the corresponding method over CC-MOGP. "+" or "-" means the result is significantly better or worse than CC-MOGP and "=" means they have similar performance.

As can be seen from the table, in terms of average accuracy, the proposed method outperforms PPSO on five datasets. The highest improvement was seen on 11Tumors with 15.54% increase on average and 12.57% in the best cases. On Leukemia2 and Brain Tumor2, the proposed method achieve similar performance to that of PPSO. On 9Tumor, the proposed method is obviously inferior to PPSO.

Compared with VLPSO, CC-MOGP have higher average accuracy on seven datasets. On 11Tumors, CC-MOGP obtained 92.3% average classification accuracy while VLPSO got 85.16% only. Only on SRBCT, CC-MOGP obtained 0.03% lower average accuracy than VLPSO.

Considering the comparison with Pareto-based methods, the average and best accuracy obtained by CC-MOGP on six datasets were significantly better than the others. Compared with PS-NSGA, the highest improvement occurred on 11Tumors with 8.4% increase on average and 10.96% higher accuracy in the best case. The obvious weakness is on Brain Tumor2 with 5.72% lower average accuracy while achieving the highest accuracy in the best case. Compared with ASMiGP, another Pareto and archive-based EA, our method is more accurate on all datasets. And the highest improvement was seen on 9Tumors with 11.52% increase on average and 13.98% in the best case.

In summary, CC-MOGP won 25, draw 5, and lost 2 out of 32 comparisons in terms of classification performance. Its results indicated that CC-MOGP conducted a much better search than the compared methods.

References

1. Zhou, Y., et al.: Imbalanced multi-fault diagnosis via improved localized feature selection. IEEE Trans. Instrum. Meas. (2023)
2. Qing, Y., et al.: Cooperative coevolutionary multiobjective genetic programming for microarray data classification. In: Proceedings of the Genetic and Evolutionary Computation Conference (2021). https://doi.org/10.1145/3449639.3459400
3. Gao, Z., Cecati, C., Ding, S.X.: A survey of fault diagnosis and fault-tolerant techniques-Part I: fault diagnosis with model-based and signal-based approaches. IEEE Trans. Ind. Electron. **62**(6), 3757–3767 (2015)
4. Dai, J., Tang, J., Huang, S., Wang, Y.: Signal-based intelligent hydraulic fault diagnosis methods: review and prospects. Chin. J. Mech. Eng. 32(1), 75 (2019)
5. Hoang, D.T., Kang, H.J.: A motor current signal-based bearing fault diagnosis using deep learning and information fusion. IEEE Trans. Inst. Meas. 69(6), 3325–3333 (2019)
6. Chen, H., Jiang, B., Ding, S.X., Huang, B.: Data-driven fault diagnosis for traction systems in high-speed trains: a survey, challenges, and perspectives. IEEE Trans. Intell. Transp. Syst. **23**(3), 1700–1716 (2022). https://doi.org/10.1109/TITS.2020.3029946
7. Yu, Y., Guo, L., Gao, H., Liu, Y.: PCWGAN-GP: a new method for imbalanced fault diagnosis of machines. IEEE Trans. Inst. Meas. **71**, 1–11 (2022). https://doi.org/10.1109/TIM.2022. 3180431
8. Zhang, Y., Li, X., Gao, L., Wang, L., Wen, L.: Imbalanced data fault diagnosis of rotating machinery using synthetic oversampling and feature learning. J. Manuf. Syst. **48**, 34–50 (2018)
9. Zhang, Y., Peng, P., Liu, C., Xu, Y., Zhang, H.: A sequential resampling approach for imbalanced batch process fault detection in semiconductor manufacturing. J. Intell. Manuf. 1–16 (2020)
10. Peng, P., Zhang, W., Zhang, Y., Xu, Y., Wang, H., Zhang, H.: Cost sensitive active learning using bidirectional gated recurrent neural networks for imbalanced fault diagnosis. Neurocomputing **407**, 232–245 (2020)

11. Liu, X.Y., Wu, J., Zhou, Z.H.: Exploratory undersampling for class-imbalance learning. IEEE Trans. Syst. Man Cybern. Part B (Cybern.) **39**(2), 539–550 (2009). https://doi.org/10.1109/TSMCB.2008.2007853
12. Chawla, N.V., Lazarevic, A., Hall, L.O., Bowyer, K.W.: SMOTEBoost: Improving prediction of the minority class in boosting. In: European Conference on Principles of Data Mining and Knowledge Discovery, pp. 107–119. Springer (2003)
13. Juez-Gil, M., Arnaiz-González, Á., Rodríguez, J.J., García-Osorio, C.: Experimental evaluation of ensemble classifiers for imbalance in Big Data. Appl. Soft Comput. **108**, 107447 (2021)
14. Bekkar, M., Alitouche, T.A.: Imbalanced data learning approaches review. Int. J. Data Min. Knowl. Manag. Process **3**(4), 15 (2013). (Academy & Industry Research Collaboration Center, AIRCC)
15. Qin, A., Mao, H., Liu, M.: Bearing fault diagnosis using modified multi-scale sample entropy and one-against-rest feature selection. In: 2021 CAA Symposium on Fault Detection, Supervision, and Safety for Technical Processes (SAFEPROCESS), pp. 1–6. IEEE (2021)
16. Mohammad-Alikhani, A., Rahnama, M., Vahedi, A.: Neighbors class solidarity feature selection for fault diagnosis of brushless generator using thermal imaging. IEEE Trans. Inst. Meas. **69**(9), 6221–6227 (2020). https://doi.org/10.1109/TIM.2020.2972081
17. Oh, H., Lee, Y., Lee, J., Joo, C., Lee, C.: Feature selection algorithm based on density and distance for fault diagnosis applied to a roll-to-roll manufacturing system. J. Comput. Des. Eng. **9**(2), 805–825 (2022). (Oxford University Press)
18. Wasikowski, M., Chen, X.: Combating the small sample class imbalance problem using feature selection. IEEE Trans. Knowl. Data Eng. **22**(10), 1388–1400 (2009). (IEEE)
19. Xu, Q., Lu, S., Jia, W., Jiang, C.: Imbalanced fault diagnosis of rotating machinery via multi-domain feature extraction and cost-sensitive learning. J. Intell. Manuf. **31**(6), 1467–1481 (2020). (Springer)
20. Ramos-Pérez, I., Arnaiz-González, Á., Rodríguez, J.J., García-Osorio, C.: When is resampling beneficial for feature selection with imbalanced wide data. Expert Syst. Appl. **188**, 116015 (2022). (Elsevier)
21. Ebrahimpour, M.K., Qian, G., Beach, A.: Multi-Head Deep Metric Learning Using Global and Local Representations. In: IEEE/CVF Winter Conference on Applications of Computer Vision (WACV), vol. 2022, pp. 1340–1349 (2022). https://doi.org/10.1109/WACV51458.2022.00141
22. Wang, C., Peng, G., Lin, W.: Robust local metric learning via least square regression regularization for scene recognition. Neurocomputing **423**, 179–189 (2021). (Elsevier)
23. Mohan, K., Seal, A., Krejcar, O., Yazidi, A.: Facial expression recognition using local gravitational force descriptor-based deep convolution neural networks. IEEE Trans. Instrum. Measur. **70**, 1–12 (2021). https://doi.org/10.1109/TIM.2020.3031835
24. Zhou, Y., Qiu, Y., Kwong, S.: Region purity-based local feature selection: a multi-objective perspective. IEEE Trans. Evol. Comput. (2022)
25. Wang, Y., Li, T.: Local feature selection based on artificial immune system for classification. Appl. Soft Comput. **87**, 105989 (2020)
26. Dudek, G.: An artificial immune system for classification with local feature selection. IEEE Trans. Evol. Comput. **16**, 847–860 (2012)
27. Armanfard, N., Reilly, J.P., Komeili, M.: Local feature selection for data classification. IEEE Trans. Pattern Anal. Mach. Intell. **38**, 1217–1227 (2015)
28. Armanfard, N., Reilly, J.P., Komeili, M.: Logistic localized modeling of the sample space for feature selection and classification. IEEE Trans. Neural Netw. Learn. Syst. **29**, 1396–1413 (2017)
29. Zhou, Y., Zhang, W., Kang, J., Zhang, X., Wang, X.: A problem-specific non-dominated sorting genetic algorithm for supervised feature selection. Inf. Sci. **547**, 841–859 (2021)
30. Zhou, Y., Kang, J., Guo, H.: Many-objective optimization of feature selection based on two-level particle cooperation. Inf. Sci. 532, 91–109 (2020). https://doi.org/10.1016/j.ins.2020.05.004
31. Zhang, Q., Li, H.: MOEA/D: a multiobjective evolutionary algorithm based on decomposition. IEEE Trans. Evol. Comput. **11**(6), 712–731 (2007)

32. Deb, K., Jain, H.: An evolutionary many-objective optimization algorithm using reference-point-based nondominated sorting approach, part I: solving problems with box constraints. IEEE Trans. Evol. Comput. **18**(4), 577–601 (2013)
33. Tran, B., Xue, B., Zhang, M.: Variable-length particle swarm optimization for feature selection on high-dimensional classification. IEEE Trans. Evol. Comput. **23**(3), 473–487 (2019)
34. Wang, P., Xue, B., Liang, J., Zhang, M.: Multiobjective differential evolution for feature selection in classification. IEEE Trans. Cybern. (2021)
35. Kennedy, J., Eberhart, R.: Particle swarm optimization. In: Proceedings of ICNN'95-International Conference on Neural Networks, vol. 4, pp. 1942–1948. IEEE (1995)
36. Kennedy, J., Eberhart, R.C.: A discrete binary version of the particle swarm algorithm. In: 1997 IEEE International Conference on Systems, Man, and Cybernetics: Computational Cybernetics and Simulation, vol. 5, pp. 4104–4108. IEEE (1997)
37. Robnik-Šikonja, M., Kononenko, I.: Theoretical and empirical analysis of ReliefF and RReliefF. Mach. Learn. **53**(1), 23–69 (2003). https://doi.org/10.1023/A:1025110714975
38. Coello, C.A., Coello, P., Toscano, G., Lechuga, M.S.: Handling multiple objectives with particle swarm optimization. IEEE Trans. Evol. Comput. **8**(3), 256–279 (2004). https://doi.org/10.1109/TEVC.2004.828267
39. Nezamabadi-pour, H., Rostami-Shahrbabaki, M., Maghfoori-Farsangi, M.: Binary particle swarm optimization: challenges and new solutions. CSI J Comput Sci Eng **6**(1), 21–32 (2008)
40. Yuan, X., Shen, S.-Q., He, Y.-L., Zhu, Q.-X.: A novel hybrid method integrating ICA-PCA with relevant vector machine for multivariate process monitoring. IEEE Trans. Control Syst. Technol. **27**(4), 1780–1787 (2019). https://doi.org/10.1109/TCST.2018.2816903
41. Tipping, M.E., Bishop, C.M.: Probabilistic principal component analysis. J. R. Stat. Soc. Ser. B (Stat. Methodol.) **61**(3), 611–622 (1999). https://doi.org/10.1111/1467-9868.00194
42. Ross Quinlan, J.: Induction of decision trees. Mach. Learn. **1**, 81–106 (1986)
43. Hastie, T., Rosset, S., Zhu, J., Zou, H.: Multi-class AdaBoost. Stat. Interface **2**(3), 349–360 (2009). https://doi.org/10.2307/44624201
44. Breiman, L.: Bagging predictors. Mach. Learn. **24**(2), 123–140 (1996). https://doi.org/10.1007/BF00994018
45. Liaw, A., Wiener, M.: Classification and regression by randomForest. R News **2**(3), 18–22 (2002)
46. Liu, Z., Cao, W., Gao, Z., Bian, J., Chen, H., Chang, Y., Liu, T.-Y.: Self-paced ensemble for highly imbalanced massive data classification. In: 2020 IEEE 36th International Conference on Data Engineering (ICDE), pp. 841–852. IEEE (2020)
47. Viola, P., Jones, M.: Fast and robust classification using asymmetric adaboost and a detector cascade. In: Advances in Neural Information Processing Systems, vol. 14 (2001)
48. Kononenko, I.: Estimating attributes: analysis and extensions of RELIEF. In: European Conference on Machine Learning, pp. 171–182 (1994)
49. Lewis, D.D., Schapire, R.E., Callan, J.P., Papka, R.: Training algorithms for linear text classifiers. In: Proceedings of the 19th Annual International ACM SIGIR Conference on Research and Development in Information Retrieval, pp. 298–306 (1996)
50. Derrac, J., García, S., Molina, D., Herrera, F.: A practical tutorial on the use of nonparametric statistical tests as a methodology for comparing evolutionary and swarm intelligence algorithms. Swarm Evol. Comput. **1**, 3–18 (2011)
51. Duggan, D.J., et al.: Expression profiling using cDNA microarrays. Nat. Genet. **21**(1), 10–14 (1999)
52. Liu, J., et al.: Transcription factor expression as a predictor of colon cancer prognosis: a machine learning practice. BMC Med. Genomics **13**, 1–10 (2020)
53. Li, J., et al.: Feature selection: a data perspective. ACM Comput. Surv. (CSUR) **50**(6), 1–45 (2017)
54. Ebrahimpour, M.K., Nezamabadi-Pour, H., Eftekhari, M.: CCFS: a cooperating coevolution technique for large scale feature selection on microarray datasets. Comput. Biol. Chem. **73**, 171–178 (2018)

55. Nag, K., Pal, N.R.: A multiobjective genetic programming-based ensemble for simultaneous feature selection and classification. IEEE Trans. Cybern. **46**(2), 499–510 (2015)
56. Poli, R., Langdon, W.B., Mcphee, N.F.: A Field Guide to Genetic Programming, vol. 10, no. 2, pp. 229–230. Lulu Press (lulu. com) (2008). (lulu. com.[SL])
57. Tran, B., Xue, B., Zhang, M.: A new representation in PSO for discretization-based feature selection. IEEE Trans. Cybern. **48**(6), 1733–1746 (2017). (IEEE)

Chapter 7
Conclusions

Abstract This chapter synthesizes the core contributions of the book into two principal themes: (1) computational intelligence (CI)-driven methodologies for feature selection and (2) their practical applications in diverse real-world scenarios. Additionally, it delineates potential avenues for future research at the intersection of CI and machine learning, emphasizing opportunities to expand their synergistic impact across broader domains.

Keywords Conclusion · Future direction

In this book, to tackle the issues and challenges of high dimensional machine learning, we mainly discussed the advanced feature selection models using CI techniques, including evolutionary algorithms based global and local FS methods, deep neural network based FS method and the case study of CI-based FS methods to solve real-world scenario problems.

Firstly, in Chaps. 3, 4 and 5, for the EA-based global FS method, PS-NSGA algorithm demonstrated superior performance in achieving higher classification accuracy and smaller feature subsets compared to other evolutionary and traditional feature selection algorithms. The accuracy-preferred non-dominated sorting and quick bit mutation were identified as crucial factors contributing to the algorithm's success; for the local FS, the exploration of a novel multi-objective Localized Feature Selection (RP-LFS) framework revealed its superiority in terms of classification accuracy and feature subset size when compared to state-of-the-art Feature Selection (FS) and Localized Feature Selection (LFS) algorithms on benchmark datasets. For the deep learning based FS method, the Hybrid Attention Pruning DNN (HAP-DNN) showcased exceptional stability and classification performance through a novel approach of sensor pruning and feature selection module. The architecture's effectiveness was validated on public human activity recognition datasets, emphasizing its potential for real-world applications.

Secondly, in Chap. 6, two real-world case studies in the field of industrial informatics and bioinformatics are conducted. On one hand, for fault diagnosis in production

© The Author(s), under exclusive license to Springer Nature Singapore Pte Ltd. 2025
Y. Zhou et al., *Computational Intelligence for High-Dimensional Machine Learning*,
SpringerBriefs in Computer Science, https://doi.org/10.1007/978-981-96-2687-8_7

equipments, the improved localized FS addressed the imbalanced multi-class challenges in fault diagnosis, achieving effective local feature subsets through a multi-objective binary PSO. The subsequent ensembles of local models enhanced the classification task's performance. On the other, the approach to multiclass microarray datasets demonstrated the utilized cooperative coevolution in a method termed CC-MOGP, which exhibited superior performance over compared methods. The integration of an index measuring complementarity between populations further enhanced the method's efficacy. The presented case studies in computational intelligence and feature selection across diverse domains showcase promising results and contribute valuable insights.

Future research avenues may involve refining the CI-based algorithms, addressing limitations, and exploring novel applications within the evolving landscape of machine learning and data-driven solutions. The success of these endeavors could further establish the practical relevance and impact of computational intelligence in real-world scenarios.